SpringerBriefs in Earth System Sciences

Series Editors

Kevin Hamilton
Gerrit Lohmann
Lawrence A. Mysak
Justus Notholt
Jorge Rabassa
Vikram Unnithan

For further volumes:
http://www.springer.com/series/10032

Diego Fernández-Prieto
Roberto Sabia

Remote Sensing Advances for Earth System Science

The ESA Changing Earth Science
Network: Projects 2009–2011

 Springer

Diego Fernández-Prieto
EO Science, Applications and Future
   Technology Department
ESA-ESRIN
Frascati
Italy

Roberto Sabia
EO Science, Applications and Future
   Technology Department
ESA-ESRIN
Frascati
Italy

ISSN 2191-589X   ISSN 2191-5903   (electronic)
ISBN 978-3-642-32520-5   ISBN 978-3-642-32521-2   (eBook)
DOI 10.1007/978-3-642-32521-2
Springer Heidelberg New York Dordrecht London

Library of Congress Control Number: 2012949381

Contents

The Changing Earth Science Network: Projects 2009–2011

Diego Fernández-Prieto, Roberto Sabia and Steffen Dransfeld

Abstract To better understand the various processes and interactions that govern the Earth system and to determine whether recent human-induced changes could ultimately de-stabilise its dynamics, both natural system variability and the consequences of human activities have to be observed and quantified. In this context, the European Space Agency (ESA) published in 2006 the document "The Changing Earth: New Scientific Challenges for ESA's Living Planet Programme" as the main driver of ESA's new Earth Observation (EO) science strategy. The document outlines 25 major scientific challenges covering all the different aspects of the Earth system, where EO technology and ESA missions may provide a key contribution. In this framework, and aiming at enhancing the ESA scientific support towards the achievement of "The Challenges", the Agency has launched the "Changing Earth Science Network", an important programmatic component of the new Support To Science Element (STSE) of the Earth Observation Envelope Programme (EOEP). In this foreword, the objectives of this initiative are summarized and the list of the projects selected in the first call, and recently completed, is provided. An in-depth overview of such projects will be provided in the following book chapters.

Keywords Earth observation · ESA · Living planet · Support to science element

D. Fernández-Prieto (✉) · R. Sabia · S. Dransfeld
European Space Agency, ESA-ESRIN, EOP Directorate,
Via GalileoGalilei s/n Frascati 00044 Rome, Italy
e-mail: diego.fernandez@esa.int

D. Fernández-Prieto and R. Sabia, *Remote Sensing Advances for Earth System Science*,
SpringerBriefs in Earth System Sciences, DOI: 10.1007/978-3-642-32521-2_1,
© The Author(s) 2013

1 ESA EO Science Strategy and the Support To Science Element (STSE)

Since their advent, satellite missions have become central in the Earth monitoring and understanding, resulting in significant progresses in a broad range of scientific areas. Although the Earth has undergone significant changes in the past, there is mounting evidence that those occurring during the last 150 years are affecting the various interactions and processes among the different components of the Earth system. Understanding those changes, their impacts on human lives and how anthropogenic activities affect the Earth system and its climate represent a major scientific endeavour where EO technology is already playing a key role.

In the mid-1990s, ESA set up its Living Planet Programme (LPP) working in close cooperation with the international scientific community to define, develop and operate focused satellite missions addressing some of the key questions at the core of Earth system science.

Moreover, realising the importance of further understanding the Earth and its response to these recent changes, the European Space Agency published "The Changing Earth: New Scientific Challenges for ESA's Living Planet Programme" as the main driver of ESA's new EO science strategy. The document outlines 25 major scientific challenges faced today covering all the different aspects of the Earth System and climate (Oceans, Atmosphere, Cryosphere, Land Surface, Solid Earth), where EO technology and ESA missions may provide a key contribution, namely:

The Challenges of the Oceans

1. Quantify the interaction between variability in ocean dynamics, thermohaline circulation, sea level, and climate.
2. Understand physical and bio-chemical air/sea interaction processes.
3. Understand internal waves and the mesoscale in the ocean, its relevance for heat and energy transport and its influence on primary productivity.
4. Quantify marine-ecosystem variability, and its natural and anthropogenic physical, biological and geochemical forcing.
5. Understand land/ocean interactions in terms of natural and anthropogenic forcing.
6. Provide reliable model- and data-based assessments and predictions of the past, present and future state of the ocean.

The Challenges of the Atmosphere

1. Understand and quantify the natural variability and the human-induced changes in the Earth's climate system.
2. Understand, model and forecast atmospheric composition and air quality on adequate temporaland spatial scales, using ground-based and satellite data.
3. Better quantify the physical processes determining the life cycle of aerosols and their interaction with clouds.

4. Observe, monitor and understand the chemistry-dynamics coupling of the stratospheric and upper tropospheric circulations, and the apparent changes in these circulations.
5. Contribute to sustainable development through interdisciplinary research on climate circulation patterns and extreme events.

The Challenges of the Cryosphere

1. Quantify the distribution of sea-ice mass and freshwater equivalent, assess the sensitivity of sea ice to climate change, and understand thermodynamic and dynamic feedbacks to the ocean and atmosphere.
2. Quantify the mass balance of grounded ice sheets, ice caps and glaciers, partition their relative contributions to global eustatic sea-level change, and understand their future sensitivity to climate change through dynamic processes.
3. Understand the role of snow and glaciers in influencing the global water cycle and regional water resources, identify links to the atmosphere, and assess likely future trends.
4. Quantify the influence of ice shelves, high-latitude river run-off and land ice melt on global thermohaline circulation, and understand the sensitivity of each of these fresh-water sources to future climate change.
5. Quantify current changes taking place in permafrost and frozen-ground regimes, understand their feedback to other components of the climate system, and evaluate their sensitivity to future climate forcing.

The Challenges of the Land Surface

1. Understand the role of terrestrial ecosystems and their interaction with other components of the Earth System for the exchange of water, carbon and energy, including the quantification of the ecological, atmospheric, chemical and anthropogenic processes that control these biochemical fluxes.
2. Understand the interactions between biological diversity, climate variability and key ecosystem characteristics and processes, such as productivity, structure, nutrient cycling, water redistribution and vulnerability.
3. Understand the pressure caused by anthropogenic dynamics on land surfaces (use of natural resources, and land-use and land-cover change) and their impact on the functioning of terrestrial ecosystems.
4. Understand the effect of land-surface status on the terrestrial carbon cycle and its dynamics by quantifying their control and feedback mechanisms for determining future trends.

The Challenges of the Solid Earth

1. Identification and quantification of physical signatures associated with volcanic and earthquake processes—from terrestrial and space-based observations.
2. Improved knowledge of physical properties and geodynamic processes in the deep interior, and their relationship to Earth-surface changes.

3. Improved understanding of mass transport and mass distribution in the other Earth System components, which will allow the separation of the individual contributions and a clearer picture of the signal due to solid-Earth processes.
4. An extended understanding of core processes based on complementary sources of information and the impact of core processes on Earth System science.
5. The role of magnetic-field changes in affecting the distribution of ionised particles in the atmosphere and their possible effects on climate.

To reinforce this strategy, in 2008 it was established the Support to Science Elements (STSE) (www.esa.int/stse), to provide scientific support for both future and on-going missions, by taking a pro-active role in the formulation of new mission concepts and products, by offering support to the scientific use of ESA EO multi-mission data and promoting the achieved results.

In this context, STSE main pillars aim at:

- Developing novel mission concepts in preparation for the next generation of European scientific missions;
- Developing advanced algorithms and innovative products that exploit the increasing ESA multi-mission capacity;
- Reinforcing ESA collaboration with the major international scientific programmes and initiatives in Earth system sciences;
- Support the Next Generation of Earth System European Scientists (The Changing Earth Science Network).

2 The Changing Earth Science Network

As one of the main programmatic components of the STSE, ESA launched in 2008 a new initiative—the Changing Earth Science Network—to support young scientists to undertake leading-edge research activities contributing to achieve the 25 scientific challenges of the LPP by maximising the use of ESA data.

The initiative is implemented through a number of research projects proposed and led by early-stage scientists at post-doctoral level for a period of 2 years. Projects undertake innovative research activities furthering into the most pressing issues of the Earth system, while exploiting ESA missions data with special attention to the ESA data archives and the new Earth Explorer missions.

Specifically, the initiative aims at:

- Contributing to the scientific advancement in Member States towards the achievement of the new 25 strategic challenges of the Living Planet Programme;
- Fostering the use of ESA EO data by the Earth Science community maximising the scientific return (in terms of scientific results and publications) of ESA EO missions;
- Contributing to consolidate a critical mass of young scientists in Europe with a good scientific and operative knowledge of ESA EO missions, assets and programmes;

Fig. 1 Distribution of the participants at the Changing Earth Science Network

- Promoting the development of a dynamic research network in ESA Member States addressing key areas of relevance for ESA missions and the ESA science strategy;
- Enhancing interactions, exchanging know-how and allowing cross fertilisation between ESA and Earth science laboratories, research centres and universities.

The first call for proposals, issued in 2008 and implemented in 2009, resulted in the selection of 11 post-doctoral scientists from the Agency's Member States based on the scientific merit of the individual projects. The final results of this call, discussed in the following chapters, include several important advances and insights in the use of ESA EO data to address some of the key current Earth science open points. A second call for proposals was issued in early 2010 to be implemented between 2011 and 2013, resulting in a further selection of 10 leading-edge research activities, which are currently on-going. A new call is in preparation and will be issued in early 2012. The map below shows the geographic distribution of the individual researchers' institutes (Fig. 1).

This volume collects some of the results obtained by the first set of projects started in 2009 and completed in 2011 (table below provide a complete list of these projects). They describe research activities exploiting data coming from several remote sensors onboard a wide suite of ESA (ERS-1/2 and Envisat), EUMETSAT, NASA and JAXA satellites (among others). The authors remarked the envisaged enhanced capabilities that will be offered in the near future with the launch of the Sentinel satellites of the GMES programme.

Acronym	Full project title	Researcher	Institute
ASSOCO	Assimilation of ocean colour satellite data to monitor the biogeochemical state of oceans and estimate its variability	Maeva Doron	Laboratoire des Ecoulements Géophysiques et Industriels, MEOM-LEGI, Grenoble, France
CARBONGASES	Retrieval and analysis of CARBON dioxide and methane greenhouse GAses from SCIAMACHY on Envisat	Oliver Schneising	Institute of Environmental Physics (IUP), University of Bremen, Bremen, Germany
CHOCOLATE	CH_4, H_2O and CO from Limb middle-ATmospher Emissions	Maya García-Comas	Instituto de Astrofísica de Andalucía (CSIC), Granada, Spain
CLARIFI	CLouds and Aerosol Radiative Interaction and Forcing Investigation: the semi-direct effect	Martin de Graaf	Royal Netherlands Meteorological Institute (KNMI), De Bilt, The Netherlands
DECPHY	Global ocean analysis of decadal covariability in phytoplankton and physical forcings through satellite remote sensing, in situ measurements and upper ocean modelling	Elodie Martinez	Laboratoire d'Océanographie de Villefranche (LOV-OOV), Villefranche sur Mer, France
DIMITRI	DIagnostics of MIxing and TRansport in atmospheric Interfaces	Elisa Palazzi	Istituto di Scienze dell'Atmosferae del Clima (ISAC-CNR), Bologna, Italy
FEMM	Fire Effects Modelling and Mapping	Patricia Oliva-Pavón	Department of Geography, University of Alcalá, Alcalá de Henares, Spain
INCUSAR	INverting consistent surface CUrrent fields from SAR	Knut-Frode Dagestad	Nansen Environmental and Remote Sensing Center, Bergen, Norway
ISMER	InSAR Survey of the Magmatic Effects on Rift development	Juliet Biggs	University of Bristol, Bristol, UK
OCCUR	Study of the chemistry-climate coupling in the UTLS region with satellite measurements	Enzo Papandrea	Department of Physics and Inorganic Chemistry, University of Bologna, Bologna, Italy
OC-FLUX	Open ocean and Coastal CO2 fluxes from Envisat and Sentinel-3 in support of global carbon cycle monitoring	Jamie Shutler	Plymouth Marine Laboratory, Plymouth, UK

In summary, the projects described in the following provide cutting-edge advanced exploitation of satellite data relevant to a broad range of scientific applications, towards an improved monitoring of the integrated Earth system.

3 Project List

The previous table provides a complete list of the projects.

Acknowledgments The authors of this Preface wish to thank ESA EOP Directorate and relevant departments/divisions for supporting the—Changing Earth Science Network—programme, and especially colleagues acting as internal contact point that have contributed to this with their expertise and dedication.

CARBONGASES—Retrieval and Analysis of Carbon Dioxide and Methane Greenhouse Gases from SCIAMACHY on Envisat

Oliver Schneising

Abstract CARBONGASES aims at making a contribution to fill the significant gaps in our understanding of the global carbon cycle by improving our knowledge about the regional sources and sinks of the two most important anthropogenic greenhouse gases, carbon dioxide and methane. To this end, global multi-year satellite data sets, namely column-averaged dry air mole fractions of carbon dioxide and methane retrieved from the SCIAMACHY instrument onboard the European environmental satellite Envisat are generated. CARBONGASES embodies seven years (2003–2009) of greenhouse gas information derived from European Earth observation data improving and extending pre-existing retrievals to maximise the quantitative information on regional surface fluxes of greenhouse gases which can be inferred from the SCIAMACHY data products using inverse modelling.

Keywords CO_2 · CH_4 · Greenhouse gases sinks and sources · Multi-year data sets

1 Introduction

Carbon dioxide (CO_2) and methane (CH_4) are the two most important anthropogenic greenhouse gases and contribute to global climate change. Both gases have increased significantly since the start of the Industrial Revolution and are now about 40 and 150 %, respectively, higher compared to the pre-industrial levels

O. Schneising (✉)
Institute of Environmental Physics (IUP), University of Bremen, FB1, Otto-Hahn-Allee 1,
P. O. Box 33 04 40 Bremen D-28334, Germany
e-mail: Oliver.Schneising@iup.physik.uni-bremen.de

D. Fernández-Prieto and R. Sabia, *Remote Sensing Advances for Earth System Science*,
SpringerBriefs in Earth System Sciences, DOI: 10.1007/978-3-642-32521-2_2,
© The Author(s) 2013

[25]. While the CO_2 concentrations have risen steadily during the last decades, atmospheric CH_4 levels were rather stable from 1999 to 2006 [3, 9] before a renewed growth was observed from surface flask measurements since 2007 [10, 18]. Despite their importance, there are still many gaps in our understanding of the sources and sinks of these greenhouse gases [26] and their biogeochemical feedbacks and response in a changing climate, hampering reliable climate predictions. However, theoretical studies have shown that satellite measurements have the potential to significantly reduce surface flux uncertainties by deducing strength and spatiotemporal distribution of the sources and sinks via inverse modelling, if the satellite data are accurate and precise enough [11, 15, 17]. The reduction of regional flux uncertainties requires high sensitivity to near-surface greenhouse gas concentration changes because the variability due to regional sources and sinks is largest in the lowest atmospheric layers.

The SCIAMACHY instrument onboard the European environmental satellite Envisat (launched in 2002) [4, 8] was the first and is now with TANSO onboard GOSAT (launched in 2009) [28] one of only two satellite instrument currently in space enabling the retrieval of carbon dioxide and methane with significant sensitivity in the boundary layer by using measurements of reflected solar radiation in the near-infrared/shortwave-infrared (NIR/SWIR) spectral range. Therefore, SCIAMACHY plays a pioneering role in the relatively new area of greenhouse gas observations from space. OCO-2 (to be launched in 2013) [2] will be another satellite designed to observe atmospheric CO_2 in the same spectral region as SCIAMACHY and TANSO. CarbonSat [5], which is one of two candidates Earth Explorer Opportunity Missions (EE-8) (to be launched in 2018) shall also measure CO_2 and CH_4 in this spectral range. The CARBONGASES project ensures that the era of continuous greenhouse gas observations from space with high sensitivity to near-surface concentration changes starts with the European Envisat satellite.

2 Methodology

The column-averaged dry air mole fractions of carbon dioxide and methane (denoted XCO_2 and XCH_4) are derived using O_2, CO_2, and CH_4 vertical colums retrieved from SCIAMACHY nadir spectra of reflected solar radiation in the near-infrared/shortwave-infrared (NIR/SWIR) spectral region with an improved version of the Weighting Function Modified DOAS (WFM-DOAS) algorithm [6, 7, 21–24]. The mole fractions are obtained from the vertical column amounts of the greenhouse gases by normalising with the air column, which can be determined by a simultaneously measured gas with less variability, e.g., O_2.

WFM-DOAS is a least-squares method based on scaling pre-selected atmospheric vertical profiles. The logarithm of the sun-normalised radiance can be assumed locally as a linear function of the vertical columns under the scaling assumption of the vertical profiles of the absorbing gases if the linearisation point

$\bar{\mathbf{V}}$ is close enough to $\mathbf{V}$, where the components of vector $\mathbf{V}$, denoted V_j, are the vertical columns of all trace gases which have absorption lines in the selected fitting window. Hence, the modelled radiation is given by

$$\ln I_{\lambda_l}^{\mathrm{mod}}(\mathbf{V}, \mathbf{b}) = \ln I_{\lambda_l}^{\mathrm{mod}}(\bar{\mathbf{V}}) + \sum_{j=1}^{J} \frac{\partial \ln I_{\lambda_l}^{\mathrm{mod}}}{\partial V_j}\bigg|_{\bar{V}_j} \cdot (V_j - \bar{V}_j) + P_{\lambda_l}(\mathbf{b}) \tag{1}$$

with the center wavelength λ_l of detector pixel number l and vector of polynomial coefficients $\mathbf{b}$ of polynomial P. A derivative, also called weighting function, with respect to a vertical column refers thereby to the change of the top-of-atmosphere radiance caused by a scaling of a pre-selected absorber concentration vertical profile.

Since the number of spectral points m in the fitting window is greater than the number n of parameters to retrieve, the problem is overconstrained and a linear least-squares approach is suitable for the retrieval of the desired vertical columns. The radiative transfer model is fitted to the logarithm of the observed sun-normalised radiance I^{obs} by minimising the difference between observation and model, i.e., the Euclidean norm of fit residuum vector $\mathbf{RES}$ (with components RES_l), for all spectral points λ_l simultaneously. The least-squares WFM-DOAS equation is then

$$\sum_{l=1}^{m} \left(\ln I_{\lambda_l}^{obs} - \ln I_{\lambda_l}^{mod}(\hat{\mathbf{V}}, \hat{\mathbf{b}}) \right)^2 \equiv \|\mathbf{RES}\|_2^2 \rightarrow \min. \tag{2}$$

where the model is given by (1) and the fit parameter vectors or vector components are indicated by a hat. The fit parameters are the desired trace gas vertical columns $\hat{V}_j$ and the polynomial coefficients. An additional fit parameter also used (but for simplicity omitted in the equations given above) is the shift (in Kelvin) of a pre-selected temperature profile. This fit parameter has been added in order to take the temperature dependence of the trace gas absorption cross-sections into account.

The least-squares problem can also be expressed in the following vector/matrix notation. Given a forward model by

$$\mathbf{y} = \mathbf{A}\mathbf{x} + \mathbf{e} \tag{3}$$

with m-dimensional measurement vector $\mathbf{y}$, n-dimensional state vector $\mathbf{x}$, $(m \times n)$ weighting function matrix $\mathbf{A}$, and model error $\mathbf{e}$, the most probable inference $p(\mathbf{x}|\mathbf{y})$ is obtained by minimising $\chi^2 = \|\mathbf{y} - \mathbf{A}\mathbf{x}\|_2^2$ with respect to $\mathbf{x}$. The solution is

$$\hat{\mathbf{x}} = \mathbf{C}_{\mathbf{x}}\mathbf{A}^T\mathbf{y}, \qquad \mathbf{C}_{\mathbf{x}} = (\mathbf{A}^T\mathbf{A})^{-1} \tag{4}$$

where $\mathbf{C}_{\mathbf{x}}$ is the covariance matrix of solution $\hat{\mathbf{x}}$.

In order to avoid time-consuming on-line radiative transfer simulations, a fast look-up table scheme has been implemented. The pre-computed spectral radiances and their derivatives (e.g., with respect to trace gas concentration and temperature profile changes) depend on solar zenith angle, surface elevation (pressure), surface albedo, and water vapour amount (to consider possible non-linearities caused by the high variability of atmospheric water vapour). These reference spectra are computed with the radiative transfer model SCIATRAN [20] for assumed (e.g., climatological) "mean" columns $\bar{V}$ solely depending on surface elevation. Multiple scattering is fully taken into account.

3 Results

All available SCIAMACHY spectra (Level 1b version 6 converted to Level 1c by the ESA SciaL1C tool using the standard calibration) for the years 2003–2009 have been processed using the improved retrieval algorithm WFM-DOAS version 2.0/2.1 [23, 24]. An overview of the long-term global data sets is shown in Figs. 1 and 2. Selected carbon dioxide and methane results are discussed in the following subsections.

3.1 Carbon Dioxide

The carbon dioxide mole fractions as a function of latitude and time are shown in Fig. 1 demonstrating the pronounced seasonal cycle due to growing and decaying vegetation and the steady increase of atmospheric carbon dioxide primarily caused by the burning of fossil fuels.

Growth rate and seasonal cycle. To examine the increase with time and the seasonal cycle more quantitatively, the SCIAMACHY results are compared to the CarbonTracker release 2010 assimilation system [16] based on monthly data. The CarbonTracker XCO_2 fields as used for this study have been sampled in space and time as the SCIAMACHY satellite instrument measures. The SCIAMACHY altitude sensitivity has been taken into account by applying the SCIAMACHY CO_2 column averaging kernels to the CarbonTracker CO_2 vertical profiles. The retrieved continuous increase with time is consistent with CarbonTracker. An analysis of global, hemispheric, and smaller zonal averages demonstrates that the annual mean increase agrees with the assimilation system within the error bars amounting to about 2 ppm/year, respectively.

For the Northern Hemisphere we also find good agreement of the phase of the CO_2 seasonal cycle with the model resulting in a pronounced correlation of the two data sets ($r = 0.98$). In contrast to the Northern Hemisphere, the seasonal cycle is less pronounced in the Southern Hemisphere and systematic phase differences are observed leading to a somewhat smaller correlation ($r = 0.91$) which is, nevertheless, quite high due to the observed consistent increase with time in both data sets. The discrepancy of the phases in the Southern Hemisphere can probably be ascribed to a large extent to the higher weight on ground scenes with occurrences of subvisual thin cirrus (induced by the restriction to land and the smaller land fraction compared to the Northern Hemisphere). Cirrus clouds are not

Fig. 1 Overview of the long-term global WFMDv2.1 carbon dioxide data set; shown are column-averaged dry air mole fractions as a function of latitude and time. In addition to the pronounced seasonal cycle due to growing and decaying vegetation, the steady increase of atmospheric carbon dioxide primarily caused by the burning of fossil fuels can be clearly observed

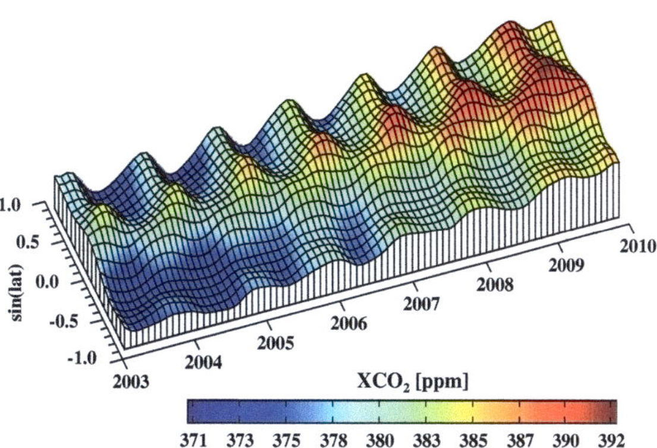

Fig. 2 Overview of the long-term global WFMDv2.0.2 methane data set; shown are column-averaged dry air mole fractions as a function of latitude and time. Clearly visible is the interhemispheric gradient and the renewed methane growth in recent years

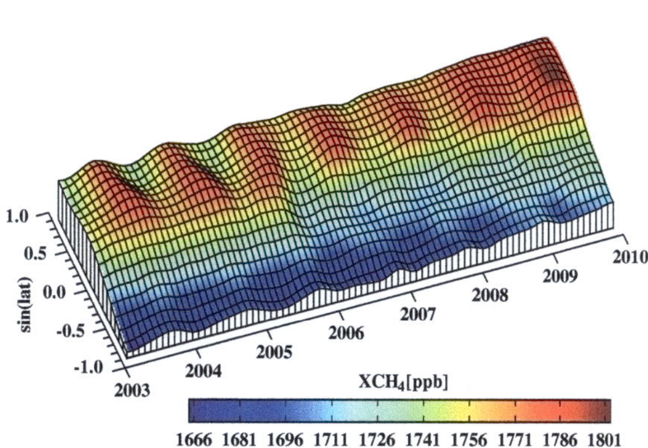

explicitly considered in WFM-DOAS yet and are hence a potential error source leading to a possible overestimation of the carbon dioxide mole fractions for scenes with high subvisual cirrus fraction [21]. Although cloud parameters are not included in the state vector of the current WFM-DOAS version, the influence of thin clouds on the retrievals is minimised resulting in much better agreement with CarbonTracker compared to the previous WFM-DOAS version especially in the Southern Hemisphere. The achieved reduction of the amplitude of the seasonal cycle is presumably a consequence of the interaction of the more realistic default aerosol scenario used in the forward model, improved spectroscopy and calibration, and the change-over to the SCIAMACHY Absorbing Aerosol Index to filter strongly aerosol contaminated scenes, in particular desert dust storms.

From the analysis of several zonal averages it can be concluded that the mean amplitude of the retrieved seasonal cycle is typically about 1 ppm larger than for CarbonTracker deriving for example 4.3 ± 0.2 ppm for the Northern Hemisphere and 1.4 ± 0.2 ppm for the Southern Hemisphere from the SCIAMACHY data.

In contrast to the growth rates, the seasonal cycle amplitude differences between SCIAMACHY and CarbonTracker are significant. The less pronounced seasonal cycle of CarbonTracker compared to the satellite data might be explainable to some extent by a too low net ecosystem exchange (NEE) between the atmosphere and the terrestrial biosphere in the underlying Carnegie-Ames Stanford Approach (CASA) biogeochemical model [13]. On the other hand, it cannot be completely excluded that undetected seasonally varying thin cirrus clouds might also contribute to some extent to the observed differences in the seasonal cycle amplitudes.

Boreal carbon uptake. Another related aspect analysed is the boreal forest carbon uptake during the growing season and its local partitioning between North America and Eurasia. To this end, longitudinal gradients of atmospheric carbon dioxide are studied during May–August (the period between the maximum and minimum of the seasonal cycle), which are the basic signals to infer regional fluxes, using a region consisting of equally sized slices in North America and Eurasia covering the bulk of the boreal forest area of the planet. When an air parcel flows over the boreal forest, more and more carbon is steadily taken up by the growing vegetation leading to a gradient parallel to wind direction with smaller values at the endpoint compared to the starting point. Due to the fact that the prevailing wind direction in mid- to high-latitudes is from west to east, one would expect a negative west-to-east longitudinal gradient for the considered region because the air masses are mainly moving according to this wind direction over the uptake region.

The gradients are derived by calculating meridional averages of seasonally averaged (May–August) SCIAMACHY and CarbonTracker XCO_2 as a function of longitude (in $0.5°$ bins) and linear fitting the corresponding west-to-east gradient weighted according to the standard deviations of the meridional averages. The associated error is derived from the square root of the covariance of the linear fit parameter. This investigation of the boreal forest carbon uptake during the growing season shows good agreement between SCIAMACHY and Carbon-Tracker concerning the annual variations of the gradients (see Fig. 3). While there is also very good quantitative agreement of the gradients for the overall region, there are systematic differences if both slices are analysed separately suggesting stronger American and weaker Eurasian uptake. However, these differences are not significant because both data sets agree within their error bars.

The suggested difference between CarbonTracker and SCIAMACHY concerning the relative strengths of the Eurasian and North American boreal forest uptake might be linked to the recent finding that modified CASA flux strengths and timings of the seasonal cycle introduce differences in corresponding gradients [12]. Therefore, a potential regional timing error in the onset of the forest uptake in the CASA model might contribute to the observed difference between CarbonTracker and SCIAMACHY. This potential contribution to the differences can be minimised by averaging over shorter time periods starting later. Actually, the restriction to June–August reduces the differences between CarbonTracker and SCIAMACHY concerning the relative regional uptake strengths to some extent, but the qualitative findings concerning regional partitioning remain the same.

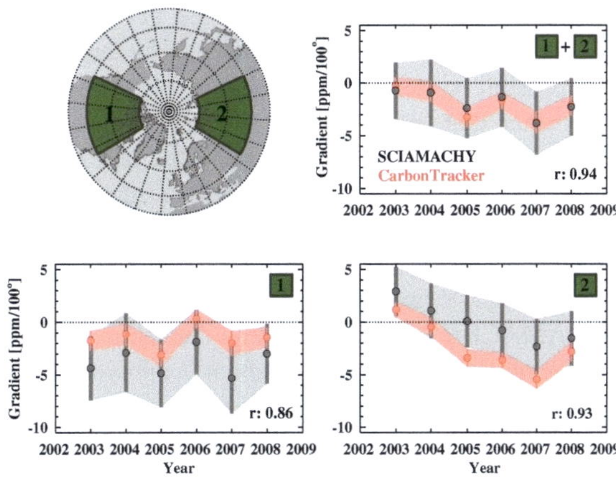

Fig. 3 Annual west-to-east longitudinal XCO$_2$ gradients from SCIAMACHY (*black*) and CarbonTracker (*red*) for boreal forests during the growing season. The examined boreal forest region is composed of two equally sized regions in North America and Eurasia. The gradients and associated errors are illustrated for the overall region and the North American and Eurasian slice separately

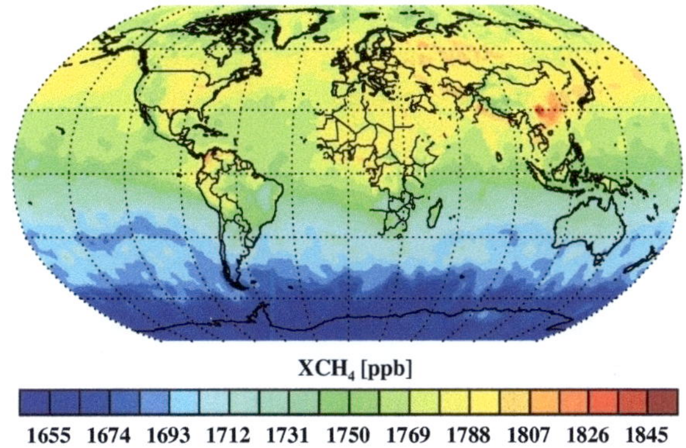

Fig. 4 Seven years mean (2003–2009) of retrieved SCIAMACHY methane. Clearly visible are major methane source regions like the Sichuan Basin in China and the interhemispheric gradient

3.2 Methane

The methane mole fractions as a function of latitude and time are shown in Fig. 2. The retrieved methane results show that after years of stability, atmospheric methane has started to rise again in recent years which is consistent with surface measurements [10, 18]. Major methane source regions like the Sichuan Basin in China which is famous for rice cultivation and the interhemispheric gradient with larger methane concentrations in the Northern Hemisphere are clearly visible in the data (see Fig. 4).

Due to proceeding detector degradation in the spectral range used for the methane column retrieval, the results since November 2005, when the impact of solar protons resulted in random telegraph noise of the detector pixel measuring the strongest CH_4 absorption in the Q-branch of the $2\nu_3$ band, are of reduced quality manifesting itself in particular by larger scatter. There are also indications for possible systematic effects; e.g, Fig. 2 suggests a potential negative tropical bias since that time.

Renewed growth. To examine the renewed methane growth in recent years more quantitatively the SCIAMACHY results are compared to the TM5-4DVAR inverse model [1, 14], which is optimised by assimilating highly accurate surface measurements at background sites from the NOAA/ESRL network, taking the SCIAMACHY averaging kernels into account. The global comparison with SCIAMACHY based on monthly data shows a pronounced correlation ($r = 0.84$) and a consistent anomaly of about 7 ppb/year since 2007.

Analysing the results for several zonal averages, the largest increase of the SCIAMACHY data is observed for the tropics and northern mid- and high-latitudes. This is consistent with [10] and the speculation that possible drivers of this renewed increase are positive anomalies of arctic temperatures and tropical precipitation.

Due care has been exercised to minimise the influence of detector degradation on the quantitative estimate of this anomaly. In this context it has to be pointed out that a static pixel mask is used since November 2005 including the entire time period of renewed increase to ensure that the observed growth is not artificially introduced by proceeding detector degradation.

3.3 Validation

From the validations with ground-based Fourier Transform Spectroscopy (FTS) measurements and comparisons with model results (CarbonTracker XCO_2 and TM5-4DVAR XCH_4) at eight Total Carbon Column Observing Network (TCCON) sites [27] distributed over Europe, America, and Australia, realistic error estimates of the satellite data are provided [24] and summarised in Table 1. Such validation is a prerequisite to assess the suitability of data sets for their use in inverse modelling. The different averaging kernels of SCIAMACHY and FTS influencing the respective absolute amounts of retrieved seasonal variability and annual increase have to be taken into account appropriately. According to [19] this

Table 1 Error characterisation of the WFM-DOAS v2.1 XCO_2 and v2.0.2 XCH_4 data products

	XCO_2 (ppm)		XCH_4 (ppb)	
	SCIA-FTS	SCIA-CT	SCIA-FTS	SCIA-TM5
Global offset	0.8	0.1	−0.3	15
Regional precision	2.2	2.2	18 (13; 16)	16 (12; 13)
Relative accuracy	1.1	1.2	19 (1.3; 21)	7.7 (2.8; 15)

The values in brackets for methane correspond to the two periods before and after the pixel mask change due to detector degradation at the beginning of November 2005 revealing the worsening of retrieval quality afterwards

can be achieved by adjusting the measurements for a common a priori profile. For simplicity, the modelled profiles (CarbonTracker for XCO_2 and TM5-4DVAR for XCH_4) are used as common a priori enabling direct comparability of SCIAM-ACHY, FTS, and the corresponding model results:

$$c_{adj} = \hat{c} + \frac{1}{p_0} \sum_l \left(1 - A^l\right) \left(x^l_{mod} - x^l_a\right) \Delta p^l \tag{5}$$

In this equation, $\hat{c}$ represents the column-averaged mole fraction retrieved by SCIAMACHY or FTS, l is the index of the vertical layer, A^l the column averaging kernel, x^l_a the a priori mole fraction, and x^l_{mod} the modelled mole fraction (and new common a priori) of layer l. p^l is the pressure difference between the upper and lower boundary of layer l and p_0 denotes surface pressure.

The relevant parameters for quality assessment are the global offset which is defined as the averaged mean difference to the reference data set over all sites, the regional precision relative to the reference which is the averaged standard deviation of the differences, and the relative accuracy which is the standard deviation of the mean differences to FTS or model simulations. The results are summarised in Table 1 indicating that SCIAMACHY carbon dioxide retrievals potentially provide valuable information for regional source/sink determination by inverse modelling techniques in places where surface flask observations are sparse and suggesting that the SCIAMACHY methane data are suitable for global inverse modelling at least before November 2005.

References

1. Bergamaschi P, Frankenberg C, Meirink JF, Krol M, Villani MG, Houweling S, Dentener F, Dlugokencky EJ, Miller JB, Gatti LV, Engel A, Levin I (2009) Inverse modeling of global and regional CH₄ emissions using SCIAMACHY satellite retrievals. J Geophys Res 114:D22301. doi: 10.1029/2009JD012287
2. Boesch H, Baker D, Connor B, Crisp D, Miller C (2011) Global characterization of CO₂ column retrievals from shortwave-infrared satellite observations of the orbiting carbon observatory-2 mission. Remote Sens 2011:270–304. doi: 10.3390/rs3020270
3. Bousquet P, Ciais P, Miller JB, Dlugokencky EJ, Hauglustaine DA, Prigent C, Van der Werf GR, Peylin P, Brunke E-G, Carouge C, Langenfelds RL, Lathiere J, Papa F, Ramonet M, Schmidt M, Steele LP, Tyler SC, White J (2006) Contribution of anthropogenic and natural sources to atmospheric methane variability. Nature 443:439–443. doi:10.1038/nature05132
4. Bovensmann H, Burrows JP, Buchwitz M, Frerick J, Noël S, Rozanov VV, Chance KV, Goede A (1999) SCIAMACHY—Mission objectives and measurement modes. J Atmos Sci 56:127–150
5. Bovensmann H, Buchwitz M, Burrows JP, Reuter M, Krings T, Gerilowski K, Schneising O, Heymann J, Tretner A, Erzinger J (2010) A remote sensing technique for global monitoring of power plant CO₂ emissions from space and related applications. Atmos Meas Tech 3:781–811. doi: 10.5194/amt-3-781-2010
6. Buchwitz M, de Beek R, Burrows JP, Bovensmann H, Warneke T, Notholt J, Meirink JF, Goede APH, Bergamaschi P, Körner S, Heimann M, Schulz A (2005) Atmospheric methane

and carbon dioxide from SCIAMACHY satellite data: initial comparison with chemistry and transport models. Atmos Chem Phys 5:941–962. doi:10.5194/acp-5-941-2005

7. Buchwitz M, de Beek R, Noël S, Burrows JP, Bovensmann H, Bremer H, Bergamaschi P, Körner S, Heimann M (2005) Carbon monoxide, methane and carbon dioxide columns retrieved from SCIAMACHY by WFM-DOAS: year 2003 initial data set. Atmos Chem Phys 5:3313–3329. doi:10.5194/acp-5-3313-2005

8. Burrows JP, Hölzle E, Goede APH, Visser H, Fricke W (1995) SCIAMACHY—Scanning imaging absorption spectrometer for atmospheric chartography. Acta Astronautica 35:445–451

9. Dlugokencky EJ, Houweling S, Masarie KA, Lang PM, Miller JB, Tans PP (2003) Atmospheric methane levels off: temporary pause or a new steady-state? Geophys Res Lett 30. doi:10.1029/2003GL018126

10. Dlugokencky EJ, Bruhwiler L, White JWC, Emmons LK, Novelli PC, Montzka SA, Masarie KA, Lang PM, Crotwell AM, Miller JB, Gatti LV (2009) Observational constraints on recent increases in the atmospheric CH_4 burden. Geophys Res Lett 36:L18803. doi: 10.1029/2009GL039780

11. Houweling S, Breon F-M, Aben I, Rödenbeck C, Gloor M, Heimann M, Ciais P (2004) Inverse modeling of CO_2 sources and sinks using satellite data: a synthetic inter-comparison of measurement techniques and their performance as a function of space and time. Atmos Chem Phys 4:523–538. doi: 10.5194/acp-4-523-2004

12. Keppel-Aleks G, Wennberg PO, Schneider T (2011) Sources of variations in total column carbon dioxide. Atmos Chem Phys 11:3581–3593. doi:10.5194/acp-11-3581-2011

13. Keppel-Aleks G, Wennberg PO, Washenfelder RA, Wunch D, Schneider T, Toon GC, Andres RJ, Blavier J-F, Connor B, Davis KJ, Desai AR, Messerschmidt J, Notholt J, Roehl CM, Sherlock V, Stephens BB, Vay SA, Wofsy SC (2011) The imprint of surface fluxes and transport on variations in total column carbon dioxide. Biogeosci. Discuss 8:7475–7524. doi:10.5194/bgd-8-7475-2011

14. Meirink JF, Bergamaschi P, Krol MC (2008) Four-dimensional variational data assimilation for inverse modelling of atmospheric methane emissions: method and comparison with synthesis inversion. Atmos Chem Phys 8:6341–6353. doi:10.5194/acp-8-6341-2008

15. Miller CE, Crisp D, DeCola PL, Olsen SC, Randerson JT, Michalak AM, Alkhaled A, Rayner P, Jacob DJ, Suntharalingam P, Jones DBA, Denning AS, Nicholls ME, Doney SC, Pawson S, Boesch H, Connor BJ, Fung IY, O'Brien D, Salawitch RJ, Sander SP, Sen B, Tans P, Toon GC, Wennberg PO, Wofsy SC, Yung YL, Law RM (2007) Precision requirements for space-based X_{CO_2} data. J Geophys Res 112:D10314. doi: 10.1029/2006JD007659

16. Peters W, Jacobson AR, Sweeney C, Andrews AE, Conway TJ, Masarie K, Miller JB, Bruhwiler LMP, Pétron G, Hirsch AI, Worthy DEJ, van der Werf GR, Randerson JT, Wennberg PO, Krol MC, Tans PP (2007) An atmospheric perspective on North American carbon dioxide exchange: CarbonTracker, Proceedings of the National Academy of Sciences (PNAS) of the United States of. America 104:18925–18930

17. Rayner PJ, O'Brien DM (2001) The utility of remotely sensed CO_2 concentration data in surface inversions. Geophys Res Lett 28:175–178

18. Rigby M, Prinn RG, Fraser PJ, Simmonds PG, Langenfelds RL, Huang J, Cunnold DM, Steele LP, Krummel PB, Weiss RF, O'Doherty S, Salameh PK, Wang HJ, Harth CM, Mühle J, Porter LW (2008) Renewed growth of atmospheric methane. Geophys Res Lett 35:L22805. doi:10.1029/2008GL036037

19. Rodgers CD (2000) Inverse methods for atmospheric sounding: theory and practice. World Scientific Publishing, Singapore

20. Rozanov VV, Buchwitz M, Eichmann K-U, de Beek R, Burrows JP (2002) SCIATRAN—a new radiative transfer model for geophysical applications in the 240–2400 nm spectral region: The pseudo-spherical version. Adv Space Res 29:1831–1835

21. Schneising O, Buchwitz M, Burrows JP, Bovensmann H, Reuter M, Notholt J, Macatangay R, Warneke T (2008) Three years of greenhouse gas column-averaged dry air mole fractions retrieved from satellite—Part 1: Carbon dioxide. Atmos Chem Phys 8:3827–3853. doi:10.5194/acp-8-3827-2008

22. Schneising O, Buchwitz M, Burrows JP, Bovensmann H, Bergamaschi P, Peters W (2009) Three years of greenhouse gas column-averaged dry air mole fractions retrieved from satellite—Part 2: Methane. Atmos Chem Phys 9:443–465. doi:10.5194/acp-9-443-2009
23. Schneising O, Buchwitz M, Reuter M, Heymann J, Bovensmann H, Burrows JP (2011) Long-term analysis of carbon dioxide and methane column-averaged mole fractions retrieved from SCIAMACHY. Atmos Chem Phys 11:2863–2880. doi:10.5194/acp-11-2863-2011
24. Schneising O, Bergamaschi P, Bovensmann H, Buchwitz M, Burrows JP, Deutscher NM, Griffith DWT, Heymann J, Macatangay R, Messerschmidt J, Notholt J, Rettinger M, Reuter M, Sussmann R, Toon GC, Velazco VA, Warneke T, Wennberg PO, Wunch D (2011) Atmospheric greenhouse gases retrieved from SCIAMACHY: Comparison to ground-based FTS measurements and model results. Atmos Chem Phys Discuss 11:28713–28745. doi:10.5194/acpd-11-28713-2011
25. Solomon S, Qin D, Manning M, Chen Z, Marquis M, Averyt KB, Tignor M, and Miller HL (eds) (2007) Climate change 2007: the physical science basis, contribution of working group I to the fourth assessment report of the intergovernmental panel on climate change (IPCC). Cambridge University Press
26. Stephens BB, Gurney KR, Tans PP, Sweeney C, Peters W, Bruhwiler L, Ciais P, Ramonet M, Bousquet P, Nakazawa T, Aoki S, Machida T, Inoue G, Vinnichenko N, Lloyd J, Jordan A, Heimann M, Shibistova O, Langenfelds RL, Steele LP, Francey RJ, Denning AS (2007) Weak northern and strong tropical land carbon uptake from vertical profiles of atmospheric CO_2. Science 316:1732–1735. doi: 10.1126/science.1137004
27. Wunch D, Toon GC, Blavier J-FL, Washenfelder RA, Notholt J, Connor BJ, Griffith DWT, Sherlock V, Wennberg PO (2011) The total carbon column observing network. Phil Trans R Soc A 369:2087–2112. doi:10.1098/rsta.2010.0240
28. Yokota T, Yoshida Y, Eguchi N, Ota Y, Tanaka T, Watanabe H, Maksyutov S (2009) Global concentrations of CO_2 and CH_4 retrieved from GOSAT: first preliminary results. SOLA 5:160–163. doi: 10.2151/sola.2009-041

CLARIFI—Aerosol Direct Radiative Effect in Cloudy Scenes Retrieved from Space–Borne Spectrometers

Martin de Graaf

Abstract The retrieval of the aerosol direct radiative effect of smoke aerosols in cloudy scenes using space–borne spectrometers is described. The retrieval of aerosol parameters and radiative effects from satellite is often hampered by residual clouds in a scene. However, aerosols that absorb solar radiation in the ultraviolet (UV) reduce the reflectance in the UV measured by space-borne spectrometers, and can be detected even in the presence of clouds. The absorption of radiation by small UV-absorbing aerosol disappears in the shortwave infrared (SWIR) and cloud properties can be retrieved here. This can be used to quantify the aerosol direct radiative effect (DRE) in the cloudy scene, by modelling the aerosol–unpolluted cloud reflectance spectrum and comparing it to the measured aerosol–polluted cloud reflectance spectrum. The algorithm to retrieve the aerosol DRE over clouds is applied here to SCIAMACHY shortwave reflectance measurements of marine cloud scenes. The maximum aerosol direct radiative effect found from these measurements is 124 ± 7 Wm^{-2}, which means that about 14 % of the incoming solar irradiance was absorbed by the smoke aerosols.

Keywords Clouds · Aerosol · Radiative transfer modelling

1 Introduction

The radiative effect of aerosols is one of the least certain components in global climate models [14]. This is mainly due to the aerosol influences on clouds. Aerosols can influence e.g. cloud formation, cloud albedo and cloud life time,

M. de Graaf (✉)
Royal Netherlands Meteorological Institute,
Wilhelminalaan 10 3732 GK De Bilt, Netherlands
e-mail: graafdem@knmi.nl

D. Fernández-Prieto and R. Sabia, *Remote Sensing Advances for Earth System Science*,
SpringerBriefs in Earth System Sciences, DOI: 10.1007/978-3-642-32521-2_3,
© The Author(s) 2013

through their role as cloud condensation nuclei, which are called the indirect effects of aerosols [8]. But even the aerosol direct radiative effect (DRE), the component of aerosol radiative forcing that neglects all influences on clouds, is still poorly constrained, due to the heterogeneous distribution of aerosol sources and sinks and the influence of clouds on global observations of aerosols. In particular, the characterisation of aerosol properties in cloudy scenes has proved challenging.

Modelling studies and observational evidence suggest that the DRE strongly depends on the underlying surface. Over dark surfaces like the ocean, the scattering effects of the aerosols dominate, leading to a negative DRE, while over bright surfaces and clouds aerosol absorption decreases the scene albedo, leading to a less negative or positive DRE [e.g.[6]]. However, modelling studies of aerosol DRE differ in magnitude and sign, because of their strong dependence on aerosol microphysical properties used in the simulations. Aerosol microphysical properties can be found from air–borne measurements, e.g. during the SAFARI 2000 field campaign [5], or globally using satellite measurements. Aerosols significantly affect the polarised light reflected by clouds under certain scattering geometries, which can be used to derive aerosol optical properties in cloudy scenes using space–borne polarimetry measurements [13]. In the case of active remote sensing, like lidar, the atmospheric scattering properties are vertically resolved, allowing for separation of aerosol and cloud properties in a small but global track [1]. The retrieved aerosol microphysical and optical properties can be used to compute the aerosol DRE over clouds, but the accuracy of these results is strongly influenced by the accuracy of the aerosol parameters that are assumed to represent the actual aerosols.

Small aerosols, like smoke from vegetation fires, reduce the scene reflectance in the UV and visible spectral region only, which may be used to retrieve the spectral optical aerosol properties in individual cases by fitting modelled reflectance spectra to the measured spectrum [3]. However, in general a unique solution is not possible, due to the large number of aerosol properties determining the reflectance spectrum. The measured reduction of UV–reflectance can also be used directly to determine the aerosol DRE in cloudy scenes, by comparing it to a reflectance spectrum of an aerosol–unpolluted scene [10]. This avoids the need for retrieved or assumed aerosol parameters. In this chapter the aerosol DRE over clouds is derived using this method with reflectance measurements of aerosol–polluted marine cloud scenes over the South Atlantic Ocean from SCIAMACHY and modelled reflectances of aerosol–unpolluted cloud scenes. The aerosol DRE over clouds can be large over the South Atlantic Ocean in the boreal summer months (June–September), when annually recurring biomass burning events during the local dry season in southern Africa produce light–absorbing aerosols that are advected over semi–permanent marine stratiform clouds [3, 11].

2 Theory

A radiative forcing or radiative effect of an atmospheric constituent x can be defined as the difference in the net irradiance ΔE at a certain level with and without the forcing constituent [7]:

$$\Delta E_x = E_{\text{with x}}^{\text{net}} - E_{\text{without x}}^{\text{net}},\tag{1}$$

where the net irradiance is defined as the difference between the downwelling and upwelling irradiances, $E^{\text{net}} = E^{\downarrow} - E^{\uparrow}$. Therefore, at the top–of–the-atmosphere (TOA), where the downwelling irradiance is the incoming solar irradiance E_0 for all scenes, the radiative effect of aerosols overlying a cloud is given by

$$\Delta E_{\text{aer}}^{\text{TOA}} = E_{\text{cld}}^{\uparrow\,\text{TOA}} - E_{\text{cld+aer}}^{\uparrow\,\text{TOA}},\tag{2}$$

where the upwelling irradiance at the TOA for an aerosol-free cloud scene is $E_{\text{cld}}^{\uparrow\,\text{TOA}}$ and the upwelling irradiance for an aerosol-polluted cloud scene $E_{\text{cld+aer}}^{\uparrow\,\text{TOA}}$. Therefore, if energy is absorbed in the atmosphere by the aerosols, the radiative forcing is positive.

The monochromatic irradiance E_λ of radiant energy is defined by the normal component of the monochromatic radiance I_λ, integrated over the entire hemisphere solid angle. In polar coordinates, this can be written as

$$E_\lambda = \frac{\mu_0 E_{0\lambda}}{\pi} \int_0^{2\pi} \int_0^1 R_\lambda(\mu, \phi; \mu_0, \phi_0)\mu\, d\mu\, d\phi.\tag{3}$$

In Eq. (3), μ_0 is the cosine of the solar zenith angle θ_0, μ the cosine of the viewing zenith angle θ, and ϕ_0 and ϕ the azimuth angle of the incoming and outgoing beam relative to the scattering plane, respectively. $\mu_0 E_0$ is the TOA solar irradiance incident on a horizontal surface unit and R is the reflectance, defined as

$$R_\lambda = \frac{\pi I_\lambda}{\mu_0 E_{0\lambda}}.\tag{4}$$

The (local) plane albedo A for a scene is defined as the integral of the reflectance R over all angles

$$A_\lambda(\mu_0) = \frac{1}{\pi} \int_0^{2\pi} \int_0^1 R_\lambda(\mu, \phi; \mu_0, \phi_0)\mu\, d\mu\, d\phi.\tag{5}$$

By substituting Eqs. (5) in (3) and integrating over wavelength, the aerosol effect at the TOA, Eq. (2), becomes

$$\Delta E_{\text{aer}} = \int_0^\infty \mu_0 E_0 \left(A_{\text{cld}} - A_{\text{cld+aer}} \right) d\lambda. \tag{6}$$

Here we have omitted the wavelength and solar zenith angle dependence of the terms on the right hand side.

The aerosol DRE over clouds can be determined using radiative transfer model (RTM) results for the first term in Eq. (6), A_{cld}, and measurements of the reflectance $R(\lambda)$ from SCIAMACHY for the second term, $A_{\text{cld+aer}}$. SCIAMACHY performs contiguous measurements from 240 and 1750 nm. Therefore, the wavelength integration is also from 240 to 1750 nm.

For the simulated case the plane albedo can be obtained from the model results, by integrating the reflectances in all directions. However, for the measured case with clouds and aerosols, only the reflectance in the measured direction is known. Therefore, the plane albedo for this scene must be estimated.

A measure for the angular distribution of the scattering energy as a function of the scattering angle for a scene is the anisotropy factor B_λ,

$$B_\lambda = R_\lambda / A_\lambda. \tag{7}$$

Assuming that the anisotropy factors are the same for the clean and polluted cloud scenes, $B_{\text{cld}} = B_{\text{cld+aer}}$, Eq. (6) can be written as

$$\Delta E_{\text{aer}} = \int_{240\text{nm}}^{1750\text{nm}} \frac{\mu_0 E_0 \left(R_{\text{cld}} - R_{\text{cld+aer}} \right)}{B_{\text{cld}}} d\lambda + \varepsilon. \tag{8}$$

The term ε contains the errors due to assumptions and measurement uncertainties. The measurement uncertainty of the aerosol DRE for SCIAMACHY was derived by applying the algorithm to aerosol–free cloud scenes. This should yield a zero aerosol DRE and differences can be attributed to systematic and random errors. Furthermore, an aerosol–polluted cloud scene was modelled using an RTM, to determine the additional errors in the algorithm from the presence of the aerosol layer. These errors were small, in the order of 1–2 Wm^{-2}. The total uncertainty of the SCIAMACHY aerosol DRE was about 7 Wm^{-2} [4].

3 Results

The aerosol DRE was derived using Eq. (8) from SCIAMACHY measurements on 13 August 2006 over the South Atlantic Ocean west of Africa. During this day smoke from biomass burning on the African mainland was drifted over the ocean in a layer between about 2–5 km altitude. Underneath this smoke layer, clouds were present over the ocean at about 1 km altitude. The horizontal distribution of the clouds and aerosols is shown by the SCIAMACHY effective Cloud Fraction in Fig. 1a and the SCIAMACHY Absorbing Aerosol Index in Fig. 1b, respectively.

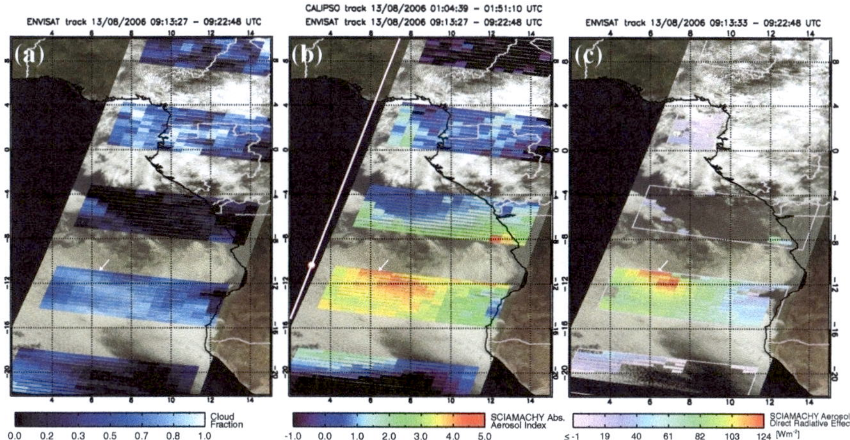

Fig. 1 MERIS RGB composite showing the horizontal cloud distribution over the west coast of Africa on 13 August 2006, from 09:13:27–09:22:48 UTC, overlaid with **a** SCIAMACHY/FRESCO effective cloud fraction; **b** SCIAMACHY Absorbing Aerosol Index; **c** SCIAMACHY Aerosol Direct Radiative Effect [Wm^{-2}], retrieved over marine clouds only. This shows the horizontal distribution of smoke over clouds over the South Atlantic Ocean, and the subsequent positive DRE due to the absorption of radiation by the aerosols. The vertical distribution of clouds and aerosols along the white Calipso track is shown in Fig. 2. The minimum distance between the Calipso track and the selected pixel (shown by the *arrow*) is 300 km. The aerosol absorption in the selected pixels is shown in Fig. 3

Fig. 2 CALIOP 532 nm backscatter signal on 13 August 2006, from 01:19:46–01:26:43 UTC, showing the vertical distribution of aerosols between 2–5 km (*yellow/green*) above clouds around 1 km (*grey*), along the Calipso track marked in white in Fig. 1. The *red arrow* corresponds to the white dot in Fig. 1

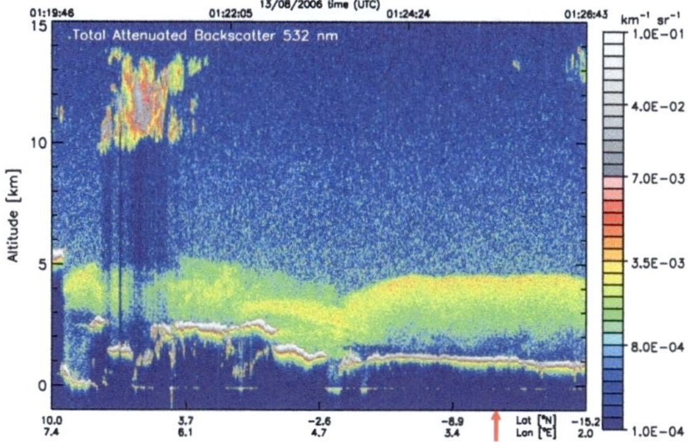

The vertical distribution of the clouds and aerosols along the Calipso track in Fig. 1b is shown in Fig. 2. The corresponding aerosol DRE field over marine clouds is shown in Fig. 1c, for all scenes over the ocean containing water clouds with effective cloud fractions greater than 0.3. It shows the unprecedented details of measured absorbed energy by aerosols over clouds. Clearly, the aerosol DRE is highly variable with location, dropping off to zero at the edges of the smoke field, which corresponds with the AAI gradient. The maximum aerosol DRE over clouds

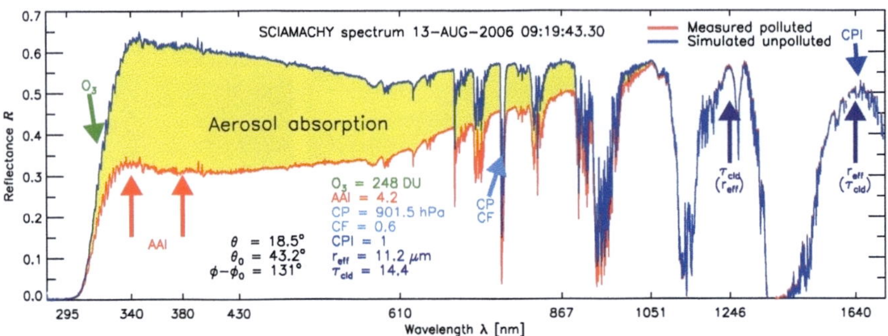

Fig. 3 SCIAMACHY measured reflectance spectrum (*red*) on 13 August 2006, 09:19:43 UTC of the scene indicated by the arrow in Fig. 1, and the modelled equivalent aerosol–unpolluted cloud reflectance spectrum (*blue*) for this scene. The difference between these two spectra (*yellow*, labelled 'aerosol absorption') indicates the irradiance absorbed by the aerosols (see Fig. 4). The parameters to model the cloud scene were retrieved at various parts of the spectrum (ozone (O_3) between 325–335 nm, cloud fraction (CF) and cloud pressure (CP) at 760 nm, cloud optical thickness (τ_{cld}) and cloud droplet effective radius (r_{eff}) at 1246 nm and 1640 nm). The AAI was retrieved from the reflectances at 340 and 380 nm

measured by SCIAMACHY on this day is 124 ± 7 Wm^{-2}, in the scene indicated by the arrow.

The measured reflectance spectrum for the scene indicated by the arrow in Fig. 1 is plotted in Fig. 3 in red. It shows the increase of the reflectance with wavelength for a scene with clouds and aerosols. The simulated reflectance spectrum of the aerosol–unpolluted cloud for this scene is plotted in blue. The cloud parameters were retrieved at various parts of the spectrum, as indicated. The cloud pressure and effective cloud fraction were derived using the oxygen-A band at 760 nm [12]. This cloud retrieval algorithm is not affected by aerosols overlying the clouds [13], if the aerosol optical thickness is reasonably small (smaller than about 1–2), which is the case for advected smoke layers. The cloud optical thickness and cloud droplet effective radius were retrieved at 1246 and 1640 nm, using simulated reflectances of water clouds [9]. At these SWIR wavelengths the aerosol absorption optical thickness is negligible and unbiased cloud parameters can be retrieved [4]. The total ozone column was retrieved using the ozone absorption between 325 and 335 nm [2]. The surface albedo (not shown) was assumed to be low and constant for the ocean. Using these scene and cloud parameters, the reflectance spectrum of an aerosol–unpolluted water cloud scene can be retrieved from pre–calculated water cloud reflectance spectrum simulations [4].

The difference between the simulated aerosol–polluted cloud scene reflectance and the measured scene reflectance is large in the UV, due to the radiation absorption by the aerosols as indicated by the yellow area. The difference disappears around 1246 nm by construction, but the measured reflectance suggests that the aerosol influence disappears already around 1100 nm.

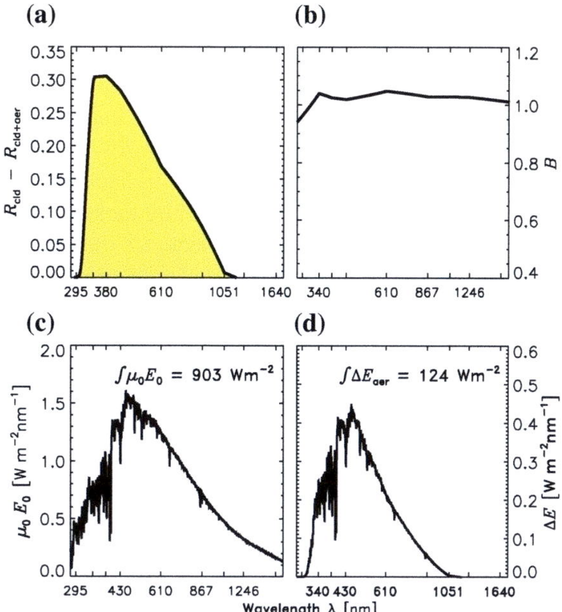

Fig. 4 The various terms of Eq. (6) to determine the aerosol DRE over clouds for the indicated scene in Fig. 1, as a function of wavelength in the SCIAMACHY spectral range. **a** The reflectance difference between the modelled aerosol–unpolluted and the measured aerosol–polluted cloud scene (same as shaded area in Fig. 3); **b** The anisotropy factor of the modelled aerosol–unpolluted cloud scene; **c** The incoming normalised solar irradiance at TOA; **d** The net irradiance change, i.e. the aerosol DRE

The various terms of Eq. (8) for the aerosol DRE in the scene indicated in Fig. 3 are illustrated in Fig. 4 as a function of wavelength in SCIAMACHY's spectral range. The reflectance difference between the simulated aerosol–unpolluted cloud reflectance spectrum and the measured reflectance spectrum $(R_{\lambda,\,\mathrm{cld}} - R_{\lambda,\,\mathrm{cld}+\mathrm{aer}})$ is given in Fig. 4a. This is the same as the yellow area in Fig. 3a. This figure clearly shows that the aerosol absorption optical thickness disappears at wavelength longer than about 1100 nm. It also disappears at wavelengths shorter than about 300 nm, because at those wavelengths all the radiation has been absorbed by ozone. This means that the SCIAMACHY spectral range suffices to capture all aerosol absorbing effects. The anisotropy factor for the modelled cloud scene $B_{\lambda,\mathrm{cld}}$ is plotted in Fig. 4b; it is typically $0.8 - 1.0$. The anisotropy factor for the aerosol–polluted cloud scene is not known, but a modelling study showed that the effect of an overlying aerosol layer on the anisotropy layer is small, at least for solar zenith angles smaller than $60°$ [4]. The normalised solar irradiance at TOA $\mu_0 E_0$ is given in Fig. 4c. The total incident solar irradiance from 240–1750 nm can be obtained by integrating the given irradiance spectrum and was 903 Wm^{-2}. The spectral irradiance change due to aerosol absorption $(E_{\lambda,\mathrm{cld}} - E_{\lambda,\mathrm{cld}+\mathrm{aer}})$ can be obtained by combining these three terms according to Eq. (6), and is plotted in Fig. 4d. By integrating over wavelength the total aerosol DRE over clouds ΔE_{aer} was found to be 124±7 Wm^{-2} for this scene.

4 Conclusions

The aerosol DRE in cloud scenes was retrieved from SCIAMACHY reflectance measurements by comparing it to modelled aerosol-unpolluted cloud reflectance spectra. The reflectance spectra of aerosol–unpolluted water clouds can be simulated using pre–computed tables of reflectances at various wavelengths. Using cloud parameters determined from the measured spectrum, at wavelengths where aerosols have no effect on the reflectance, the equivalent aerosol–unpolluted cloud reflectance spectrum for a scene can be simulated. The reflectance change in the UV due to radiation absorption by aerosols can be converted to a shortwave flux directly, which avoids the need for aerosol parameters retrievals or assumptions. This can help validate modelling results of aerosol DRE, which use the modelled radiative fluxes in an aerosol–loaded and aerosol–free scene. These results are commonly very sensitive to the assumed aerosol optical and micro-physical properties.

The SCIAMACHY measured aerosol DRE over clouds in the South Atlantic Ocean was found to be as large as 124 ± 7 Wm^{-2}, which means the aerosols absorbed about 14 % of the incoming solar radiation. The measured aerosol DRE over clouds show details due to variations in the smoke and cloud fields, that can currently not be resolved by chemistry–transport models. Therefore, the measurements from SCIAMACHY and other space–borne spectro(radio)meters may prove very valuable for understanding the radiative effects of aerosols on clouds.

The use of retrieved cloud optical thickness and cloud droplet effective radius to construct a (water cloud) reflectance spectrum implies an implicit separation of the aerosol DRE in cloudy scene from that in clear skies. This is one of the areas where observations of aerosol DRE are currently lacking [14]. Consequently, the method presented here can complement studies that retrieve aerosol parameters in clear–sky only. The latter may be used to derive the aerosol DRE in clear skies.

Acknowledgments This work was financed by the European Space Agency (ESA) within the Support to Science Element, project number 22403.

References

1. Chand D, Anderson TL, Wood R, Charlson RJ, Hu Y, Liu Z, Vaughan M (2008) Quantifying above-cloud aerosol using spaceborne lidar for improved understanding of cloud-sky direct climate forcing. J Geophys Res 113:D13206. doi:10.1029/2007JD009433
2. Eskes HJ, van der ARJ, Brinksma EJ, Veefkind JP, de Haan JF, Valks PJM (2005) Retrieval and validation of ozone columns derived from measurements of SCIAMACHY on Envisat. Atmos Chem Phys Disc 5:4429–4475. doi:10.5194/acpd-5-4429-2005
3. De Graaf M, Stammes P, Aben EAA (2007) Analysis of reflectance spectra of UV-absorbing aerosol scenes measured by SCIAMACHY. J Geophys Res 112:D02206. doi:10.1029/2006JD007249

4. De Graaf M, Tilstra LG, Stammes P (2011) Retrieval of the aerosol direct radiative effect over clouds from space-borne spectrometry. J Geophys Res

5. Haywood JM, Osborne SR, Francis PN, Neil A, Formenti P, Andreae MO, Kaye PH (2003) The mean physical and optical properties of regional haze dominated by biomass burning aerosol measured from the C-130 aircraft during SAFARI 2000. J Geophys Res 108:D13. doi:10.1029/2002JD002226

6. Keil A, Haywood JM (2003) Solar radiative forcing by biomass burning aerosol particles during SAFARI 2000: a case study based on measured aerosol and cloud properties. J Geophys Res 108:D13. doi:10.1029/2002JD002315

7. Liou KN (2002) An introduction to atmospheric radiation. Academic Press, Amsterdam

8. Lohmann U, Feichter J (2005) Global indirect aerosol effects: a review. Atmos Chem Phys 5 doi:1680-7324/acp/2005-5-715

9. Nakajima T, King MD (1990) Determination of the optical thickness and effective particle radius of clouds from reflected solar radiation measurements: Part I Theor J Atmos Sci 47(15):1878–1893

10. Stammes P, Tilstra LG, Braak R, de Graaf M, Aben EAA (2008) Estimate of solar radiative forcing by polluted clouds using OMI and SCIAMACHY satellite data. In: Proceedings of the international radiation symposium (IRS2008), pp. 577–580. Foz do Iguaçu, Brazil

11. Swap RJ, Garstang M, Macko SA, Tyson PD, Maenhaut W, Artaxo P, Källberg P, Talbot R (1996) The long-range transport of southern African aerosols to the tropical South Atlantic. J Geophys Res 101:D19. doi:10.1029/95JD01049

12. Wang P, Stammes P, van der AR, Pinardi G, van Roozendael M (2008) FRESCO+: an improved O_2 A-band cloud retrieval algorithm for tropospheric trace gas retrievals. Atmos Chem Phys 8(21):6565–6576. doi:10.5194/acp-8-6565-2008

13. Waquet F, Riedi J, Labonnote L, Goloub P, Cairns B, Deuzé J-L, Tanré D (2009) Aerosol remote sensing over clouds using A-train observations. J Atmos Sci 66. doi:10.1175/2009JAS3026.1

14. Yu H, Kaufman YJ, Chin M, Feingold G, Remer LA, Anderson TL, Balkanski Y, Bellouin N, Boucher O, Christopher S, DeCola P, Kahn R, Koch D, Loeb N, Reddy MS, Schulz M, Takemura T, Zhou M (2006) A review of measurement-based assessments of the aerosol direct radiative effect and forcing. Atmos Chem Phys 6(3):613–666. doi:10.5194/acp-6-613-2006

DIMITRI—Diagnostics of Mixing and Transport in Atmospheric Interfaces

Elisa Palazzi

Abstract Knowledge of mass exchange between different parts of the atmosphere is essential for understanding the distribution of atmospheric trace constituents, such as ozone, water vapour, methane, nitrous oxide, and aerosols. It is also important for understanding long- and short-term changes of key phenomena, such the springtime ozone hole over the Antarctic stratosphere. The project DIMITRI—Diagnostics of Mixing and Transport in atmospheric Interfaces (June 2009–June 2011)—was focused on the analysis of mass exchange between the tropics and the extra-tropics, across the so-called subtropical barrier. The study of transport processes across the subtropical barrier was approached from a tracer perspective, that is, through the analysis of long-lived tracer concentrations measured from satellite and their statistical properties. The methodology employed is based on the construction of the Probability Distribution Functions (PDFs) of tracer concentrations, from which the position of the subtropical barrier can be derived. It was assessed to which extent the characteristics of the four different satellite instruments employed (MIPAS/ Envisat, MLS/Aura, SMR/Odin, HALOE/UARS) affect the representation of the subtropical barrier, in order to draw conclusions from the synopsis of data. All sensors coherently reproduce the annual cycle and interannual variability of the subtropical barrier. However, some differences occur in the representation of the Quasi Biennial Oscillation (QBO) effects on the subtropical barrier position, probably due to the uneven sampling of QBO westerly and easterly periods in the different data sets. A 18 years long time series (from 1992 up to now) of the subtropical barrier position has been built from the merging of the four satellite data sets employed, which will be used in future studies focused on the detection of

E. Palazzi (✉)
National Research Council (CNR), Institute of the Atmospheric Sciences
and Climate (ISAC), Turin, Italy
e-mail: E.Palazzi@isac.cnr.it

D. Fernández-Prieto and R. Sabia, *Remote Sensing Advances for Earth System Science*, SpringerBriefs in Earth System Sciences, DOI: 10.1007/978-3-642-32521-2_4,

possible trends reflecting natural or human-induced changes in stratospheric transport and circulation.

Keywords Troposphere · Stratosphere · Transport processes · Mixing ratio

1 Introduction

The meridional circulation of the stratosphere, known as the Brewer Dobson circulation, is a global-scale cell in which air rises in the tropics and moves poleward and downward, mostly in the winter hemisphere, and it is now generally accepted as the basic description of the seasonal-mean meridional mass transport in the stratosphere. The addition of quasi-horizontal mixing is the main refinement to a "simple" picture of a stratospheric mass transport from low to high latitudes driven, as in the troposphere, by thermal forcing. Isentropic mixing and diabatic overturning are driven by the same mechanism: breaking of synoptic- and planetary-scale waves from the troposphere in the stratosphere. The region in the mid-latitude stratosphere where planetary-scale waves break down, generating stirring and irreversible mixing, is called "surf-zone". The mid-latitude surf zone is separated from the tropics by a dynamical barrier to transport, the subtropical barrier, where steep meridional gradients in long-lived tracer concentrations are found.

Both aspects of stratospheric transport, vertical advection and isentropic mixing, affect the distribution of long-lived atmospheric tracers. Vertical advection acts to steepen isopleths in the barrier regions while isentropic mixing acts to flatten isopleths in the stirring regions, yet concurring in the production of very strong gradients at the stirring region edge. Improvements in the understanding of stratospheric circulation and transport across dynamical barriers have progressed with advances in understanding of stratospheric dynamics, with accumulating satellite observations of the distribution of tracers, and with the development and refinement of specific diagnostic tools. Despite that, important key questions still remain unresolved, which requires both experimental and modeling efforts. The project DIMITRI, focused on the analysis of transport processes across the subtropical barrier, aimed at addressing the specific challenges of the ESA Living Planet Program (http://www.esa.int/esaLP/), such as the observation and understanding of the chemistry-dynamics coupling of the stratospheric and upper tropospheric circulations and the apparent changes in these circulations, and the quantification of the natural variability and the human-induced changes in the Earth's climate system. The specific objectives of the project were to (1) define a set of appropriate diagnostics to make an effective use of atmospheric chemical composition data from satellite measurements; (2) characterize transport mechanisms in the lower and middle stratosphere across the subtropical barrier; and (3) start a process-oriented evaluation in the subtropical barrier region of the state-of-the-art chemistry climate models (CCMs).

2 Methodology and Data

Analysis of passive tracers is among the best tools to study transport in the stratosphere as given by the combination of vertical advection and isentropic mixing. The methodology employed within DIMITRI to highlight some aspects of transport across the subtropical barrier is based on the analysis of the PDFs of tracer concentrations and their statistical properties (e.g., [4, 9]). Tracer PDFs are defined as the likelihood (frequency) of observing a tracer value χ in the range from $\chi - \Delta\chi$ and $\chi + \Delta\chi$, in a given region of the atmosphere, normalized by the total area of the region.

All long-lived tracer PDFs are similar under similar dynamic conditions: in the winter hemisphere they are characterized by three modes (PDF maxima) and two troughs (PDF minima), while in the summer hemisphere by two modes separated by one region of minimum values in the tracer PDF. A clear mapping can be established between the structural features (maxima and minima) of the tracer PDF and the tracer versus latitude scatter plot. This is clearly shown in the example of Fig. 1. Panel A shows the scatter plot of nitrous oxide (N2O) volume mixing ratio (VMR) concentrations (y-axis) versus latitude at 580 K (~ 30 hPa) during January 2004, measured by the Michelson Interferometer for Passive Atmospheric Sounding (MIPAS) instrument onboard Envisat. Panel B (C) shows the N2O PDF versus the tracer VMR concentrations, turned sideways, in the Southern Hemisphere (Northern Hemisphere). Peaks in the PDFs map into flat regions in the scatter plot (tracer values in the tropics, surf-zone, polar vortex interior), while minima in the PDFs map into steeply sloping regions in the scatter plot (tracer values in the subtropical barrier and polar vortex edge regions). The MIPAS N2O vertical profiles were generated at the Institute for Meteorology and Climate Research (IMK) through the IMK-IAA data processor [11].

The mapping between the tracer PDF structure and tracer versus latitude scatter plot can be exploited to associate a single characteristic latitude to the subtropical barrier (and polar vortex edge) position, making use of the statistics of the "support region". From a mathematical point of view, to describe the statistic of the support the notation of conditional PDF is adopted. The conditional PDF $P(\phi|\chi)$ is the likelihood of the latitudes ϕ of measurements having mixing ratios in a neighbourhood of the tracer value χ. As an example, Fig. 1 shows that the subtropical minimum in the January 2004 PDF occurs at a mixing ratio χ^* (called "tracer boundary" of the subtropics) of about 250 ppbv in both the NH and the SH (red horizontal lines). Mapping the range of values around χ^* across from the PDF (separately for each hemisphere) to the scatter plot along the horizontal lines and then down to the latitude axis in panel D, locates the support of the tracer boundary of the subtropics, that is, the region in the latitude space over which the tracer field takes on values near $\chi^* = 250$ ppbv. Panel D shows the conditional PDF $P(\phi|\chi^*)$, for the two subtropical regions and the vortex edge region and the maximum value of $P(\phi|\chi^*)$ for each of these regions occur at one single latitude corresponding to

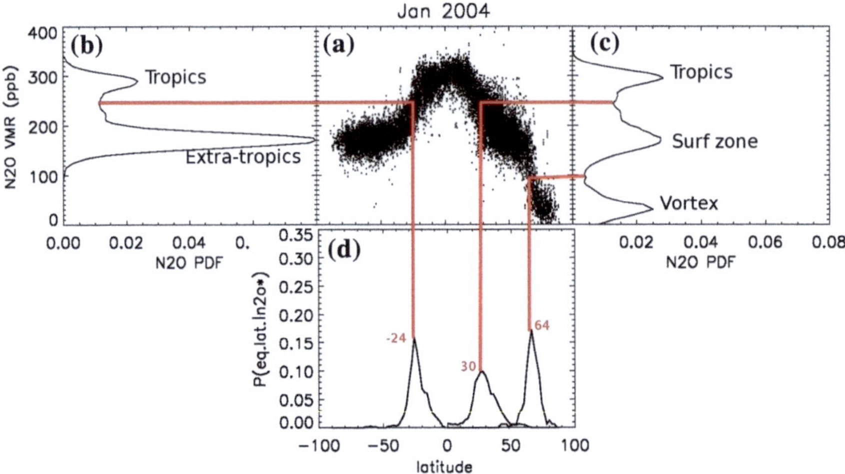

Fig. 1 Panel A: scatter plot of MIPAS N2O VMR concentrations (y-axis) versus latitude at 580 K (January 2004). Panel C (B): N2O PDF versus the tracer VMR concentration, turned sideways, in the NH (SH). Panel D: PDF of the support in the subtropical regions and vortex edge region (see text)

the subtropical barrier (30° N and 24° S in this example) and the polar vortex barrier (64° N in this example).

Within DIMITRI, the aforementioned methodology was applied to N2O and methane (CH4) satellite measurements. There are currently four satellite instruments in orbit measuring N2O. Three of these were used within DIMITRI and are MIPAS onboard Envisat [1], the microwave limb sounder (MLS) onboard Aura [2, 12], and the sub-millimeter radiometer (SMR) onboard Odin [3, 10]. CH4 data were taken from the HALOE/UARS instrument [7], which stopped sounding the atmosphere on 2005. For more details on data products, the instrument characteristics and the most important differences among the sensors having a role for the study reported on here, see Palazzi et al. [5] and references therein.

3 Results: The Subtropical Barrier from Satellite Data

N2O PDFs from MLS/Aura (2004–2009), MIPAS/Envisat and SMR/Odin (2002–2009), and CH4 PDFs from HALOE/UARS (1992–2005) have been calculated on a seasonal rather than shorter basis (e.g., monthly or daily), to overcome the temporal sampling inhomogeneities among the different satellite instruments. A careful comparison among the different satellites of the morphological features of one- and two-dimensional tracer PDFs is presented in Palazzi et al. [5].

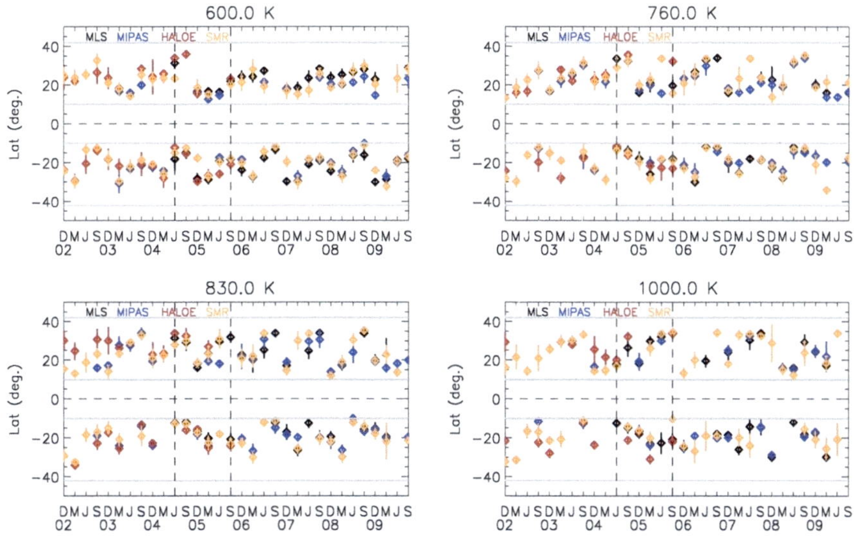

Fig. 2 Seasonal time series of the subtropical barrier position from DJF 2002 to SON 2009 calculated from MLS (*black*), MIPAS (*blue*), HALOE (*red*), and SMR (*yellow*) data. *Error bars are superimposed to each data point*

Figure 2 shows the time series of the subtropical barrier position from 2002 to 2009 (though the HALOE data set extends back till 1992) at four potential temperature levels (600, 760, 830 and 900 K), calculated from MLS/MIPAS/SMR N2O and HALOE CH4 PDFs. An overall good agreement among the four satellite instruments can be noticed. The systematic differences among the sensors have been carefully evaluated and discussed in Palazzi et al. [5]: in summary, they do not exceed 2° in the overlapping time period among all satellite sensors (2004–2005, dashed vertical lines in Fig. 2).

The four sensors consistently reproduce the annual cycle of the subtropical barrier position, characterized by a wintertime shift of the barrier position toward the equator. The time series plotted in Fig. 2 (particularly at 600 K and 690 K) also clearly show a periodicity of approximately 2 years in the position of the subtropical barrier, which is ascribed to the impact of the quasi biennial oscillation (QBO) in the zonal mean zonal winds, the main source of interannual variability in the lower stratosphere. Using the monthly mean zonal wind components measured at Singapore (1° N, 104° E) from 1992 to 2009, the effect of the QBO on the subtropical barrier position calculated from the four data sets was evaluated; the results are summarized in Fig. 3.

For each sensor, blue (red) symbols denote the average position (over all measurement years available for each sensor) of the subtropical barrier during the westerly (easterly) QBO phase. All sensors consistently reproduce the effect of the QBO on the subtropical barrier in the SH and NH at 600 K, which was identified

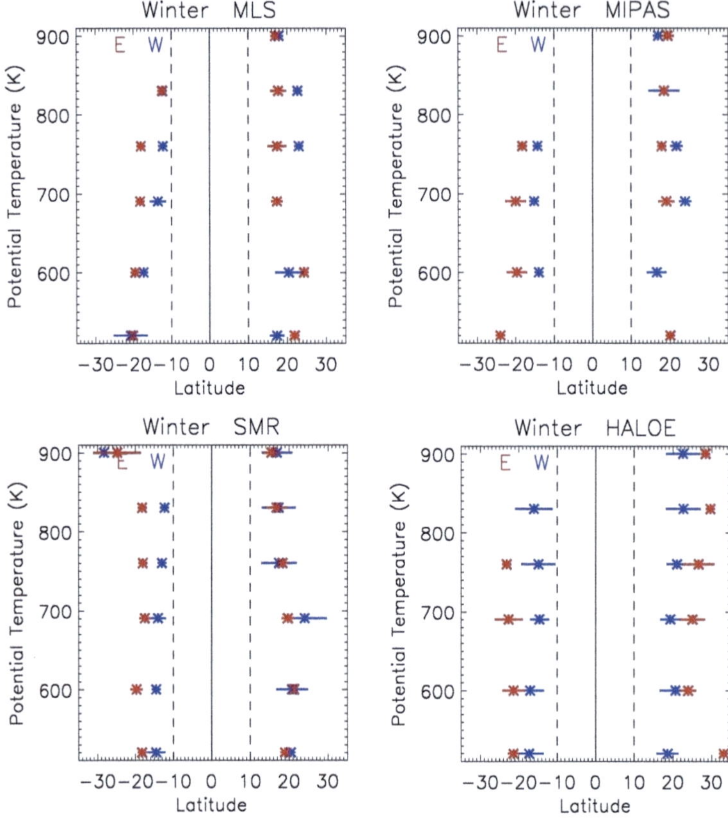

Fig. 3 Mean position (average over all years of data available for each sensor) of the subtropical barrier position for QBO Easterly (E, *red*) and Westerly (W, *blue*) phases, for MLS (*top left*), MIPAS (*top right*), SMR (*bottom left*), and HALOE (*bottom right*)

in previous studies as the level of maximum QBO influence on lower stratospheric transport, see Shuckburgh et al. [8].

The effect of the QBO at this level is to enhance the wintertime shift of the subtropical edge toward the equator (or the summer hemisphere) during the westerly QBO phase. In general, the sensors agree better in the SH than in the NH.

4 Conclusions

Much of what we know about stratospheric transport has been inferred from observations of long-lived tracers. The relatively recent abundance of constituent observations from satellite has provided an excellent opportunity to study transport as reflected by changes in tracer distributions. Such observations, for instance, have shown that transport and mixing in the stratosphere are highly inhomogeneous, with regions of strong stirring separated by barrier regions (such as the subtropical barrier) across which there is relatively little or even nonexistent horizontal transport, see Plumb [6]. The study carried out within DIMITRI was focused on the subtropical barrier, which controls isentropic exchanges between the tropics and the extra-tropics. Two long-lived tracers, N2O and CH4, measured by four different satellite instruments (MIPAS/Envisat, MLS/Aura, SMR/Odin, and HALOE/UARS), were analyzed with a common species-independent diagnostics (tracer PDF) to supply a quantitative assessment of the subtropical barrier variability.

Despite their large differences in the latitudinal coverage, temporal and spatial sampling, the instrumental characteristics and the measured chemical species, the four sensors consistently reproduce the seasonal cycle of the subtropical barrier, characterized by the wintertime shift of the subtropical edge toward the summer hemisphere. The quasi-biennial variability of the subtropical barrier, ascribed to the effect on stratospheric transport of the QBO, is not captured by all sensor data in the same way at all levels considered, probably due to their different temporal coverage and the uneven representations of QBO westerly and easterly periods.

The research performed within DIMITRI demonstrated that the PDF approach answers to the need for synthesis of a large and growing database of observations of the atmosphere, which can be extended to the outputs generated by atmospheric models. Since models cannot capture all the details of a tracer field, there is a need to perform comparisons between observations and model data that are to some extent independent from these details, which can be done using statistical approaches like that of tracer PDFs.

Acknowledgments The DIMITRI work was carried out at the ISAC institute in Bologna under the supervision of Dr. Federico Fierli and in collaboration with Dr. Gabriele Stiller from the Karlsruhe Institute of Technology and Dr. Joachim Urban from the Gothenburg University. Elisa Palazzi acknowledges the European Space Agency (ESA) for supporting the project DIMITRI in the framework of the ESA Changing Earth Science Network Initiative. MIPAS level-1b data were provided by ESA. MLS/Aura and HALOE/UARS data were acquired as part of the activities of NASA's Science Mission Directorate, and are archived and distributed by the Goddard Earth Sciences (GES) Data and Information Services Center (DISC). Odin is a Swedish–led satellite project funded jointly by Sweden (SNSB), Canada (CSA), Finland (TEKES), France (CNES), and the third–party mission program of the European Space Agency (ESA).

References

1. Fischer H et al (2008) MIPAS: an instrument for atmospheric and climate research. Atmospheric Chem Phys 8:2151–2188. doi:10.5194/acp-8-2151-2008
2. Lambert A et al (2007) Validation of the aura microwave limb sounder middle atmosphere water vapor and nitrous oxide measurements. J Geophys Res 112:D24S36. doi:10.1029/2007JD008724

3. Murtagh D et al (2002) An overview of the Odin atmospheric mission. Can J Phys 80:309–319
4. Palazzi E, Fierli F (2010) Diagnostics of mixing and transport in the atmospheric interfaces: the project DIMITRI. Proceedings of the ESA living planet symposium, Bergen, Norway, 28 June–2 July 2010, ESA SP-686
5. Palazzi E, Fierli F, Stiller GP, Urban J (2011) Probability density functions of long-lived tracer observations from satellite in the subtropical barrier region: data intercomparison. Atmospheric Chem Phys 11:10579–10598. doi:10.5194/acp-11-10579-2011
6. Plumb RA (2002) Stratospheric transport. J Meteorol Soc Jpn 80:793–809
7. Russell JM III, Gordley LL, Park JH, Drayson SR, Hesketh WD, Cicerone RJ, Tuck AF, Frederick JE, Harries JE, Crutzen PJ (1993) The halogen occultation experiment. J Geophys Res 98:10777–10797
8. Shuckburgh E, Norton W, Iwi A, Haynes P (2001) The influence of the quasi-bienniel oscillation on isentropic transport and mixing in the tropics and subtropics. J Geophys Res 106:14327–14338
9. Sparling LC (2000) Statistical perspectives on stratospheric transport. Rev Geophys 38: 417–436
10. Urban J et al (2005) Odin/SMR limb observations of stratospheric trace gases: validation of N2O. J Geophys Res. 110:D09301. doi:10.1029/2004JD005394
11. von Clarmann T, et al (2009) Retrieval of temperature, H2O, O3, HNO3, CH4, N2O, ClONO2 and ClO from MIPAS reduced resolution nominal mode limb emission measurements. Atmospheric Meas Tech 2:159–175. doi:10.5194/amt-2-159-2009
12. Waters JW et al (2006) The earth observing system microwave limb sounder (EOS MLS) on the aura satellite. IEEE T Geosci Remote 44:1075–1092

OCCUR—Study of the Chemistry−Climate Coupling in the UTLS Region with Satellite Measurements

Enzo Papandrea and Massimo Carlotti

Abstract This study, performed in the frame of the Changing Earth Science Network initiative, aims at investigating in particular the Upper Troposphere-Lower Stratosphere (UTLS) region with measurements acquired from satellite instruments and the aid of a chemical transport model. It is related to the "Challenge 4 of the Atmosphere" of the ESA Living Planet Program. The Challenge 4 addresses the following topic: Observe, monitor and understand the chemistry-dynamics coupling of the stratospheric and upper tropospheric circulations, and the apparent changes in these circulations. To contribute to this scientific challenge this study investigated the stratosphere and the UTLS region through the analysis of satellite observations using a tomographic approach, capable to take into account inhomogeneities of the atmosphere's constituents. In this context, evidence of non-physical day-night differences in 1D retrievals has been found in a previous study performed on MIPAS spectra. These differences almost disappear using a tomographic approach, confirming that a 2D retrieval describes a more realistic picture of the atmosphere. These evidences have been confirmed with a 2D retrieval algorithm that has been expressly set up to analyse SCIAMACHY observations. The results of the study have also provided a new scientific understanding and support the definition of requirements for future Earth Explorers missions (i.e. an advanced imaging MIPAS spectrometer such as PREMIER).

Keywords Troposphere · Stratosphere · Ozone · Chemical transport model

E. Papandrea (✉) · M. Carlotti
Dipartimento di Chimica Fisica e Inorganica, Università di Bologna,
Viale del Risorgimento, 4 40136 Bologna, Italy
e-mail: enzo.papandrea@unibo.it

D. Fernández-Prieto and R. Sabia, *Remote Sensing Advances for Earth System Science*, SpringerBriefs in Earth System Sciences, DOI: 10.1007/978-3-642-32521-2_5,
© The Author(s) 2013

1 Introduction

A good understanding of the Upper Troposphere/Lower Stratosphere (UTLS) region is crucial as dynamical, chemical, and radiative coupling between the stratosphere and troposphere are among the many important processes that must be understood for prediction of global changes [12].

Transport in this region is a key factor: the Brewer-Dobson circulation determines the distribution of stratospheric ozone, while in the UTLS air mass exchange is driven by upward transport at tropical latitudes and downward transport at high latitudes.

There is evidence that anthropogenic emission of ozone-depleting substances (ODSs) caused substantial depletion of stratospheric ozone [17]. With the adoption of the Montreal Protocol (1987) and subsequent amendments, the global production of ODSs was brought under control and strongly reduced. As a consequence, the ozone concentration in the stratosphere, and in particular the so called Antarctic ozone hole, is expected to recover to 1980s values around 2060–2070 [9, 18]. However, the influence of increased greenhouse gases and natural variability cause large uncertainties on projections of ozone recovery.

In order to better understand these phenomena this study is focused on the analysis of limb measurements of two instruments that are both onboard the Envisat satellite.

Although the short term variability and inhomogeneity of the atmosphere contain a signature of the processes governing the atmosphere, to date most of the retrievals of remotely-sensed atmospheric measurements are still carried out with one-dimensional (1D) algorithms see e.g. [15, 23, 24]. These algorithms make the assumption that the atmosphere is horizontally homogeneous along the lines of sight of the observations. Two-dimensional (2D) tomographic retrieval algorithms (see e.g. [2, 16, 21, 26]) allow to take into account the inhomogeneity of the atmosphere in the along-track direction of limb-measurements and disclose to the exploitation of more information from the measurements themselves.

The necessity of a more realistic description of inhomogeneous atmospheric fields has been addressed in a study performed on MIPAS observations. This study [14] reported that non-physical day-night differences are present in 1D retrievals from MIPAS spectra, mostly at locations where the horizontal temperature gradients of the atmosphere are strong. These differences almost disappear using a tomographic approach, suggesting that a 2D retrieval describes a more realistic picture of the atmosphere. These considerations generally act as a support to all the work performed in this study. Here, a tomographic approach analysis for the measurements of two different instruments (MIPAS and SCIAMACHY) both onboard the Envisat satellite, is described. To improve our understanding of the atmosphere, future Earth Explorers missions with improved capabilities are necessary. In this view, the analysis of a new hypothetical instrument, operating in the infrared spectral region as MIPAS but having a different spectral resolution/ observation geometry e.g. PREMIER [13], has been performed.

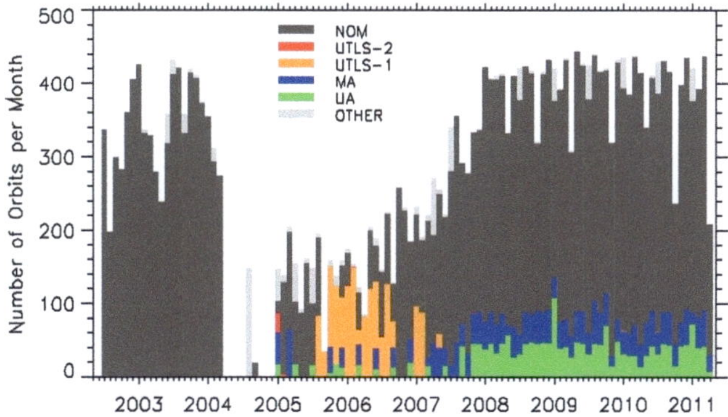

Fig. 1 Distribution of the number of orbits per month for the main MIPAS observation modes (from the beginning of the mission up to 19 April 2011)

2 Production, Updating, and Maintenance of GMTR Level 2 Database of MIPAS Full Mission

MIPAS is a limb-scanning Fourier transform (FT) spectrometer recording emission spectra in the mid-infrared, with a spectral range extending from 680 to 2,410 cm^{-1} [11]. The original full resolution (FR) mission was suspended on March 26 2004, due to the deterioration of the interferometer slides. This technical problem was overcome in January 2005 by operating MIPAS with an optimized resolution (OR) at 41 % of the original spectral resolution.

Two-dimensional fields of MIPAS observations have been obtained through the GMTR analysis code that is based on the Geo-fit approach [3] along with the multi-target tetrieval (MTR) functionality [7]. The code performs a tomographic retrieval on observations collected along a whole orbit operating a 2-D discretization of the atmosphere thus enabling the horizontal atmospheric structures to be modelled. In this approach each limb observation contributes to determine the unknown quantity at a number of different locations among those spanned by its line of sight. The MTR functionality enables to perform the simultaneous retrieval of different geophysical parameters.

Using the GMTR code we have recently obtained a Level 2 database. It is named MIPAS2D, covers the time frame March 2002 to April 2011 and contains 2D fields of pressure, temperature and volume mixing ratio (VMR) of key atmospheric constituents (H_2O, O_3, HNO_3, CH_4, N_2O, NO_2, N_2O_5, $ClONO_2$, CFC-11 and CFC-12) retrieved from MIPAS measurements [8, 19].

Figure 1 gives an updated overview of the Level 1b orbits used in the database for the MIPAS nominal (NOM), middle atmosphere (MA) and UTLS-1 observation modes. The retrieved fields have been filtered on the basis of the χ^2_R values and other quality quantifiers [8]. They are made available to the scientific community.

As an example of atmospheric field contained in the MIPAS2D database, Fig. 2 shows 1-month averaged temperature values computed at a fixed pressure level

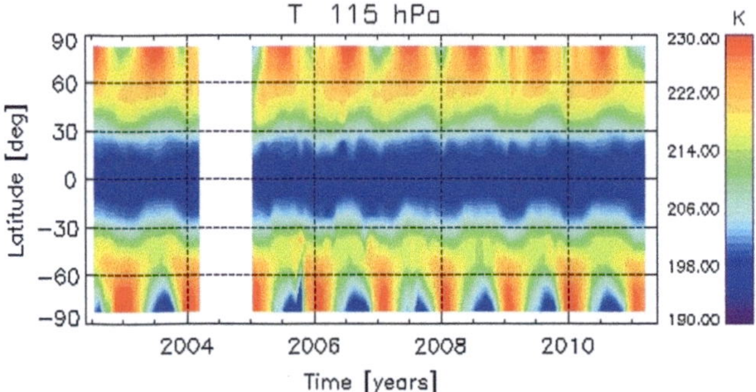

Fig. 2 Temperature monthly averaged at 115 hPa pressure level. The latitude grid is separated by 15°. MIPAS NOM and UTLS-1 observations have been merged

(115 hPa, corresponding to 16.8 km) and for a latitude grid binned at steps of 15°. The map clearly shows the seasonality: cold winters and warm summers are observed in the polar regions. As expected, the lowermost temperatures correspond to the southern polar vortex. This map highlights the consistence between FR and OR results. An exhaustive work that compares 1D/2D retrievals from MIPAS spectra has been published in Kiefer et al. [14].

In order to have a fully homogeneous database for the whole period, a degradation of the resolution of all the spectra of the FR MIPAS mission has been performed. This analysis consists of a subsequent retrieval of the atmospheric fields from the degraded spectra using the auxiliary data of the OR mission. In this dataset the OR get significantly closer to the FR, showing that the FR-OR bias (that is observed for some species mostly depending on altitude) is mainly due to systematic errors introduced by the different sets of microwindows (MWs) and auxiliary data adopted for the analysis of the two missions (e.g. [6]). The MIPAS2D database has been positively compared with the chemical transport model BASCOE [10].

3 Development of a Retrieval Algorithm Capable to Analyze SCIAMACHY Measurements

SCIAMACHY is a passive imaging spectrometer that measures the solar scattered radiation in the UV–VIS–NIR spectral ranges from 240 to 2,380 nm with a wavelength depending spectral resolution ranging from 0.2 to 1.5 nm [1]. In the nominal mode, the instrument sounds the atmosphere alternating sequences of nadir and limb measurements. Limb-scans are performed looking along the flight direction with an approximate vertical sampling of 3.3 km. The SCIAMACHY instrument differs therefore from MIPAS mainly because it can acquire also nadir measurements and it can measure only during daytime.

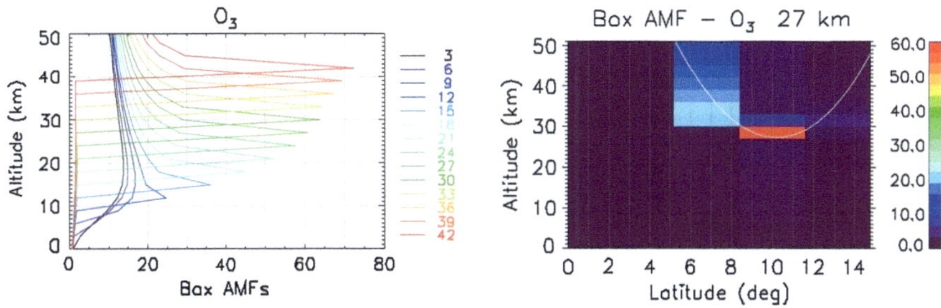

Fig. 3 O_3 1D Box AMFs (*left*) for tangent heights from 3 to 42 km and 2D box AMFs for 27 km tangent height (*right*)

We have explored the possibility to adopt a tomographic approach also for SCIAMACHY measurements, going in the same direction followed with MIPAS data.

The procedure consists in averaging over all spectra at each elevation step in a given limb scan obtaining measures suitable for a 2D algorithm. The spectra are subsequently used as inputs for a DOAS spectral fitting program and the radiative transfer model (RTM). In this spectral range the RTM (unlike MIPAS) must take into account multiple scattering. Afterwards we generate slant column densities (SCDs) for each limb-scanning sequence at each considered tangent height for the absorber of interest. Finally, we solve the inverse problem for SCDs to obtain the vertical profile using box airmass factors (box-AMFs) generated by the 3D Montecarlo RTM code MOCRA [20].

The box-AMFs (which are used to convert SCDs of trace gases into vertical column densities) describe the dependence of the measured SCDs with respect to changes in concentration of the considered boxes, therefore they measure the spatial sensitivity of the measurements to the gas present in the boxes in which the atmosphere is stratified.

Figure 3 reports, in the left-hand panel, an example of 1D AMFs (depending only on altitude) for tangent heights from 3 to 42 km and, in the right-hand panel, 2D box AMFs (depending both on altitude and latitude). The map of the 2D case shows for clarity only the viewing geometry having the tangent height at 27 km. The additional latitudinal dimension of the 2D case, representing a more realistic stratification of the atmosphere, enhances the value of the box AMFs around the tangent point. Both in the 1D and in the 2D cases the implemented inversion formula makes use of a priori information to stabilize the retrieval described in Rodgers [25].

Using the developed codes we then selected some test cases to account for the difference between 1D and 2D algorithms both with simulated and real observations. A reference atmosphere characterized by strong horizontal gradients has been used to test the capability of the tomographic algorithm to reproduce the horizontal inhomogeneities. Ozone SCDs have been simulated through the MOCRA code that has been upgraded for this purpose. The SCDs have been subsequently used both in the 1D and in the 2D version of the retrieval code.

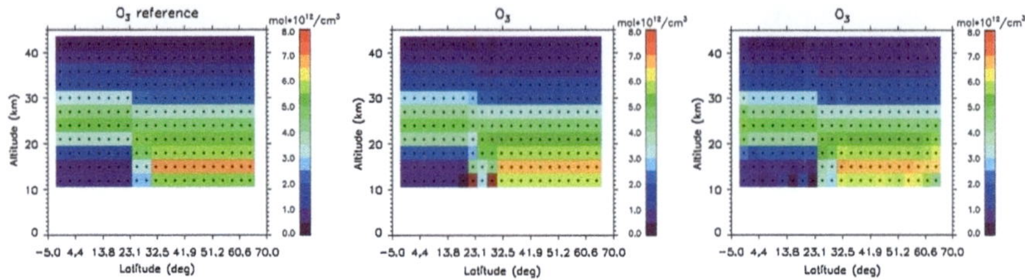

Fig. 4 Reference atmosphere for ozone (*left*), 1D retrieval (*middle*), 2D retrieval (*right*)

The outcome of this analysis is shown in Fig. 4 where the reference ozone number density (left), the 1D retrieval (middle) and the 2D retrieval (right) are reported. The figure shows that both codes correctly reproduce portions of the atmosphere where the number density varies smoothly in the horizontal domain. On the contrary, when a steep variation in ozone number density is present, the 1D retrieval cannot represent properly these regions of the atmosphere. The 2D version is instead able to properly reproduce also these structures.

Using real measurements, in many regions of the atmosphere, where horizontal gradients are present, differences between the 1D and 2D has been found by Pukite et al. [22].

We also performed some tests using the suitable "limb only" orbits acquired on 14 December 2008, confirming that the two approaches lead to a difference that in many cases is greater than the retrieval error. Indeed, since the measure is more sensitive to the side between the tangent point and the instrument, when a negative gradient is present, the lower number density field closer to the instrument dominates the observed SCD (leading to a lower number density in the 1D retrieval), while in presence of a positive gradient the opposite happens. These results are evidenced in Fig. 5 where NO_2 number density for orbit 35499 retrieved by the 1D and 2D approach (top) are shown together with the 1D/2D differences (center), and with the 1D random errors (bottom).

4 Sensitivity of Different Sensors to the UTLS Region

We investigated the sensitivity of the UTLS region to the measurements of two instruments (MIPAS and an "advanced MIPAS") operating with different observation geometries through developed tools. The hypothetical advanced instrument is an imaging spectrometer able to sound the atmosphere with a very fine scan-pattern that would permit innovative studies, e.g. the structure of the UTLS with high level of detail. In our analysis we considered this instrument having a signal to noise requirement allowed by modern array detector technologies and operating about 400 limb-scanning sequences/orbit, each sounding the atmosphere at 25 tangent altitudes separated by 2 km from 6 to 54 km. For this spectrometer a set of microwindows has been selected for the retrieval of ozone and temperature.

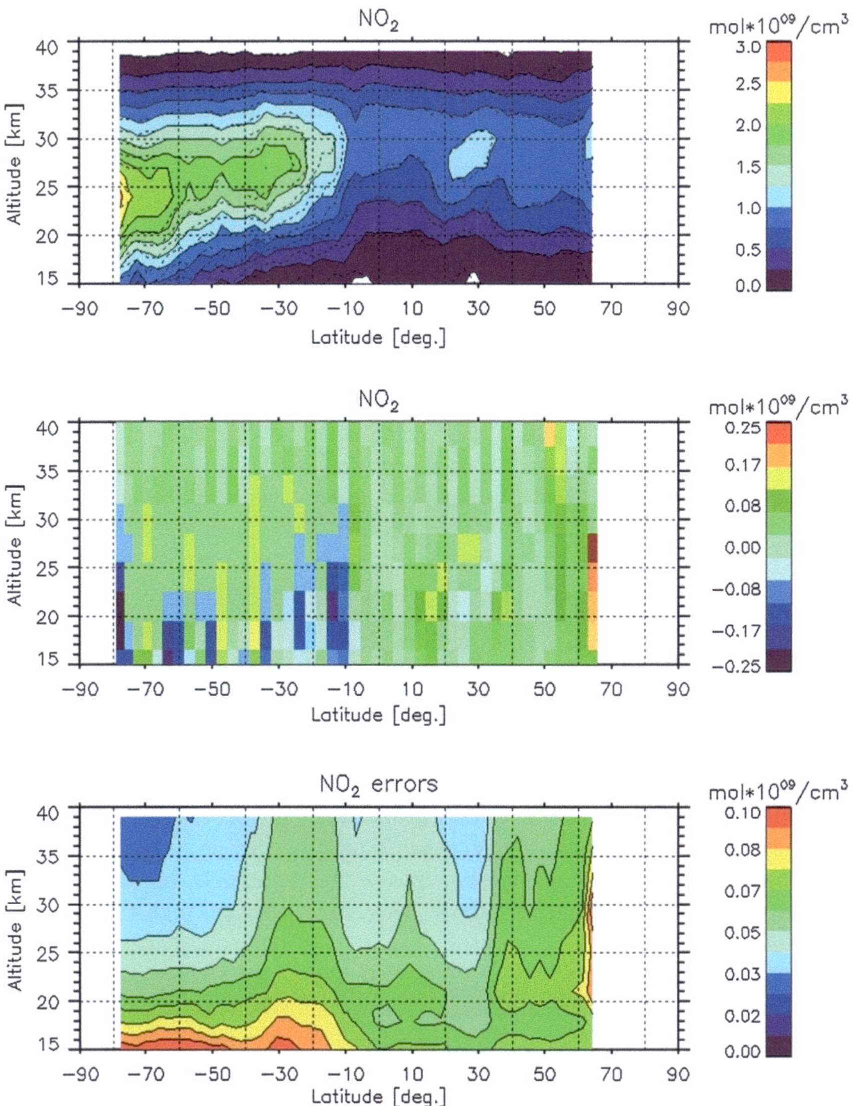

Fig. 5 *Top* NO$_2$ number density for orbit 35499 retrieved by the 2D approach (*color plot with solid contourlines*) and 1D approach (*dashed contourlines*). *Center* absolute differences between the 1D and 2D approach. *Bottom* random errors of the 2D case

The selection has been operated on the basis of the error and the 'Information Load (IL)' [4] analysis.

For MIPAS, its three observation modes that measure the UTLS region have been investigated, in particular with respect to the NOM, UTLS-1, UTLS-2 observation modes performances in terms of strength, spatial coverage and

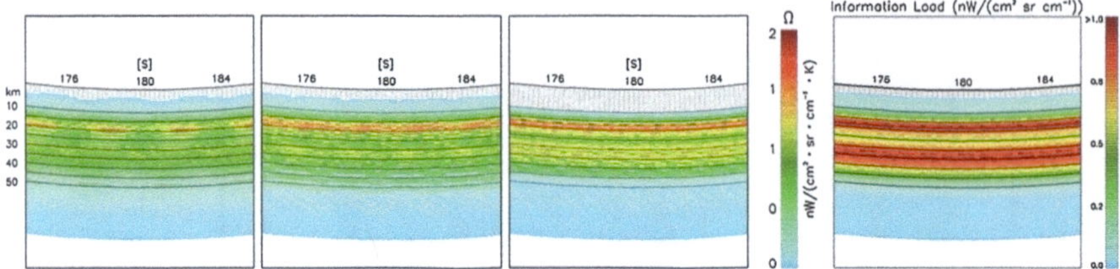

Fig. 6 Information load (IL) distributions with respect to ozone around the south pole. From *left* to *right*: MIPAS NOM, UTLS-1, UTLS-2, "advanced MIPAS". The vertical dimension of the atmosphere is expanded by a factor of ten with respect to the extension of the earth's radius

uniformity of the IL distribution along the full orbit [5]. This approach is only possible using a two-dimensional representation of the atmospheric fields.

This study showed that the two observation modes that were specifically designed for the UTLS region are actually competitive with the third one, that was designed for the whole stratosphere, up to altitudes that far exceed the UTLS. In the UTLS the performance of the two specific observation modes is comparable even if the best performance in terms of horizontal resolution is provided by the UTLS-2 observation mode that currently is excluded from MIPAS observation planning.

This analysis has been extended here using the advanced MIPAS characteristics and an overview of the IL distributions is given in Fig. 6 where the four panels show the IL distributions with respect to ozone around the South Pole in the case of (from left to right): MIPAS NOM, UTLS-1, UTLS-2 and an advanced MIPAS.

From a visual inspection it is clearly evident that the advanced instrument would have the capability to extract much more information, particularly in the UTLS region. Furthermore its very fine measurement grid determines a horizontal and vertical uniformity of the sensitivity. This characteristic is highly desirable, as it makes easier the interpretation of the results. The retrieval study on simulated observations is in progress.

5 Summary and Conclusions

The OCCUR project has been focused on a tomographic retrieval approach for different instruments aimed to study the UTLS region through limb measurements acquired from satellite instruments. A 2D analysis system has been applied to the infrared MIPAS and the UV/VIS SCIAMACHY both onboard Envisat.

The necessity of a more realistic description of inhomogeneous atmospheric fields has been evidenced in several occasions.

Through the GMTR analysis code recently a Level 2 database (MIPAS2D) has been obtained. It covers the time frame March 2002 to April 2011 and contains 2D fields of pressure, temperature and VMR of key atmospheric constituents.

A 2D retrieval code has been set up also for the SCIAMACHY observations. Differences both on simulated spectra and on real orbits have been showed to be greater than the retrieval error when strong horizontal gradients are present.

In order to improve the picture of our atmosphere (especially of the UTLS region that is characterized by high variability and several interconnected phenomena), future missions having improved characteristics would be welcomed. To this purpose, in this study we have considered a new instrument exploiting imaging technology, that has characteristics of the PREMIER mission. The analysis shows that, in terms of precision and spatial resolution, it would allow a big step forward in the knowledge of several atmospheric aspects (a manuscript is in preparation on this subject).

Acknowledgments E. Papandrea acknowledges support by ESA within the framework of the Changing Earth Science Network Initiative.

References

1. Bovensmann H, Burrows JP, Buchwitz M, Frerick J, Noël S, Rozanov VV, Chance KV, Goede APH (1999) SCIAMACHY: mission objectives and measurement modes. J Atmos Sci 56:127–150
2. Carlotti M, Dinelli BM, Raspollini P, Ridolfi M (2001) Geo-fit approach to the analysis of limb-scanning satellite measurements. Appl Opt 40:1872–1885
3. Carlotti M, Brizzi G, Papandrea E, Prevedelli M, Ridolfi M, Dinelli BM, Magnani L (2006) Two-dimensional geo-fit multitarget retrieval model for Michelson interferometer for passive atmospheric sounding/environmental satellite observations. Appl Opt 45:716–727
4. Carlotti M, Magnani L (2009) Two-dimensional sensitivity analysis of MIPAS observations. Opt Express 17:5340–5357
5. Carlotti M, Papandrea E, Castelli E (2011) Two-dimensional performance of MIPAS observation modes in the upper-troposphere/lower-stratosphere. Atmos Meas Tech 4: 355–365. doi:10.5194/amt-4-1-2011
6. Ceccherini S, Cortesi U, Verronen PT, Kyrölä E (2008) Technical note: continuity of MIPAS-Envisat operational ozone data quality from full- to reduced-spectral-resolution operation mode. Atmos Chem Phys 8:2201–2212
7. Dinelli BM, Alpaslan D, Carlotti M, Magnani L, Ridolfi M (2004) Multi-target retrieval (MTR): the simultaneous retrieval of pressure, temperature and volume mixing ratio from limb-scanning atmospheric measurements. J Quant Spec Radiat Transfer 84:141–157
8. Dinelli BM, Arnone E, Brizzi G, Carlotti M, Castelli E, Magnani L, Papandrea E, Prevedelli M, Ridolfi M (2010) The MIPAS2D database of MIPAS/Envisat measurements retrieved with a multi-target 2-dimensional tomographic approach. Atmos Meas Tech 3:355–374
9. Eyring V et al (2007) Multimodel projections of stratospheric ozone in the 21st century. Geophys Res Lett 112:D16303
10. Errera Q, Daerden F, Chabrillat S, Lambert JC, Lahoz WA, Viscardy S, Bonjean S, Fonteyn D (2009) 4D-var assimilation of MIPAS chemical observations: ozone and nitrogen dioxide analyses. Atmos Chem Phys 8:6169–6187
11. Fischer H, Birk M, Blom C, Carli B, Carlotti M, von Clarmann T, Delbouille L, Dudhia A, Ehhalt D, Endemann M, Flaud JM, Gessner R, Kleinert A, Koopman R, Langen R, López-Puertas M, Mosner P, Nett H, Oelhaf H, Perron G, Remedios J, Ridolfi M, Stiller G, Zander R (2008) MIPAS: an instrument for atmospheric and climate research. Atmos Chem Phys 8:2151–2188

12. Holton JR, HaynesPH, Douglass AR, Rood RB, Pfister L (1995) Stratosphere–troposphere exchange. Rev Geophys 33:403–439
13. Kerridge BG, Barnett J, Birk M, Buehler S, Vandaele A-C, Carlotti M et al (2005) Process exploration through measurements of infrared and millimetre-wave emitted radiation (PREMIER). Proposal for ESA core explorer mission: CCLRC, SSTD, SSTD-RSG
14. Kiefer M, Arnone E, Dudhia A, Carlotti M, Castelli E, von Clarmann T, Dinelli BM, Kleinert A, Linden A, Milz M, Papandrea E, Stiller G (2010) Impact of temperature field inhomogeneities on the retrieval of atmospheric species from MIPAS IR limb emission spectra. Atmos Meas Tech Discuss 3:1707–1742
15. Kuhl S, Pukıte J, Deutschmann T, Platt U, Wagner T (2008) SCIAMACHY limb measurements of NO_2, BrO and OClO. Retrieval of vertical profiles: algorithm, first results, sensitivity and comparison studies. Adv Space Res 42:1747–1764
16. Livesey NJ, Van Snyder W, Read WG, Wagner PA (2006) Retrieval algorithms for the EOS microwave limb sounder (MLS). IEEE T Geosci Remote 44:1144–1155
17. Molina MJ, Rowland FS (1974) Stratospheric sink for chlorofluoromethanes: chlorine atomic-catalysed destruction of ozone. Nature 249:810–812
18. Newman PA et al (2006) When will the Antarctic ozone hole recover? Geophys Res Lett 33:L12814
19. Papandrea E, Arnone E, Brizzi G, Carlotti M, Castelli E, Dinelli BM, Ridolfi M (2010) Two-dimensional tomographic retrieval of MIPAS/Envisat measurements of ozone and related species. Int J Remote Sens 31(2):477–483
20. Premuda M, Masieri S, Bortoli D, Margelli F, Ravegnani F, Petritoli A, Kostadinov I, Giovanelli G, Cupini E (2009) A Monte Carlo simulation of radiative transfer in the atmosphere applied to ToTaL-DOAS. In: Picard RH, Schäfer K, Comeron A, Kassianov EI, Mertens CJ (eds) Remote sensing of clouds and the atmosphere XIV. Proceedings of SPIE, vol 7475, SPIE, Bellingham, WA, p 74751A
21. Pukıte J, Kuhl S, Deutschmann T, Platt U, Wagner T (2008) Accounting for the effect of horizontal gradients in limb measurements of scattered sunlight. Atmos Chem Phys 8:3045–3060
22. Pukıte J, Kuhl S, Deutschmann T, Dorner S, Jockel P, Platt U, Wagner T (2010) The effect of horizontal gradients and spatial measurement resolution on the retrieval of global vertical NO_2 distributions from SCIAMACHY measurements in limb only mode. Atmos Meas Tech 3:1155–1174
23. Raspollini P, Belotti C, Burgess A, Carli B, Carlotti M, Ceccherini S, Dinelli BM, Dudhia A, Flaud J-M, Funke B, Hpfner M, L'opez-Puertas M, Payne V, Piccolo C, Remedios JJ, Ridolfi M, Spang R (2006) MIPAS level 2 operational analysis. Atmos Chem Phys 6:5605–5630
24. Ridolfi M, Carli B, Carlotti M, von Clarmann T, Dinelli BM, Dudhia A, Flaud J-M, Hoepfner M, Morris PE, Raspollini P, Stiller G, Wells RJ (2000) Optimized forward model and retrieval scheme for MIPAS near-real-time data processing. Appl Opt 39(8):1323–1340
25. Rodgers CD (2000) Inverse methods for atmospheric sounding. Theory and practice. World Scientific Publishing Co. Ltd., Singapore, p 238
26. Steck T, Hopfner M, von Clarmann T, Grabowski U (2005) Tomographic retrieval of atmospheric parameters from infrared limb emission observations. Appl Opt 44:3291–3301

DecPhy—Variability of the Phytoplankton Spring and Fall Blooms in the Northeastern Atlantic in the 1980s and 2000s

Elodie Martinez, David Antoine, Fabrizio D'Ortenzio
and Clement DeBoyerMontégut

Abstract Phytoplankton chlorophyll-a (Chl) seasonal cycles of the Northeastern Atlantic are described using satellite ocean color observations covering the 1980s and the 2000s. The study region is where warmer SST and higher Chl in the 2000s as compared to the 1980s have been reported (30°–50° N and 40°–0° W). In this area, two phytoplankton blooms take place: a spring bloom that follows stratification of upper layers, and a fall bloom due to nutrient entrainment through deepening of the mixed layer. In the 1980s, spring and fall blooms were of similar amplitude over the entire study region. In the 2000s, the fall bloom was weaker likely due to a delayed deepening of the mixed layer at the end of summer (mixed-layer depths—MLD—determined from in situ data). Conversely, the spring bloom was stronger in the 2000s than it was in the 1980s, in parallel to a deeper MLD and stronger winds in winter. Our results show that the links between upper-layer stratification, SST changes, and biological responses are more complex than the simple paradigm that sequentially relates higher stratification with warmer SST and an enhanced (weakened) growth of the phytoplankton population in the subpolar (subtropical) region.

Keywords Physical and biogeochemical interactions · Interannual variability · Phytoplankton · Physical-biological coupling · Mixed layer · Decadal oscillations

E. Martinez (✉) · D. Antoine · F. D'Ortenzio
Laboratoire d'Océanographie de Villefranche, LOV, UMR 7093,
CNRS & UPMC Univ Paris, 06230 Villefranche-sur-Mer, France
e-mail: elodie.martinez@ird.fr

E. Martinez
IRD, Laboratoire d'Océanographie Physique et Biogéochimique, Centre d'Océanologie de Marseille, Faculté des Sciences de Luminy, 13288 Marseille cedex 9, France

C. DeBoyerMontégut
Laboratoire d'Océanographie Spatiale, IFREMER Pointe du Diable, Plouzané, France

D. Fernández-Prieto and R. Sabia, *Remote Sensing Advances for Earth System Science,* 49
SpringerBriefs in Earth System Sciences, DOI: 10.1007/978-3-642-32521-2_6,
© The Author(s) 2013

1 Introduction

Phytoplankton spring blooms in the North Atlantic are the most pronounced of any open-ocean region [1]. They have been the focus of many studies [2–4]. Mechanisms of their rise were first described 60 years ago [5], and are still being investigated [4, 6–8]. The onset of a spring bloom occurs when upper-layer stratification is sufficient for phytoplankton to use the nutrients previously brought to the surface by deep winter mixing [5]. This is also the time when surface irradiance increases towards the summer solstice maximum. Spring blooms end when surface waters are nutrient-depleted due to consumption by phytoplankton (bottom-up control), and also because of zooplankton grazing (top-down control) [9].

Fall blooms have been reported in the North Atlantic Drift Region (NADR; 43–56° N; 43–0° W) [7, 10]. This bloom in fall is conversely driven by the deepening of the surface mixed layer at the end of summer, leading to nutrient entrainment in the surface layer. It is also controlled by a reduction in grazing pressure due to the seasonal vertical migration of mesozooplankton towards deep waters [9]. It is dissipated (diluted) by mixing and energy-starved by declining light in fall-winter [10].

The aim of this study is to investigate differences in the Northeastern Atlantic (30°–50° N and 40°–0° W) phytoplankton blooms in spring and fall between 1979–1983 and 1998–2002. These two time periods are respectively the first 5 years of the Coastal Zone Color Scanner (CZCS, NASA ocean color mission launched in 1978 [11]), and Sea-viewing Wide Field-of-view Sensor (SeaWiFS, NASA mission launched in 1997 [12]), satellite ocean color missions. The surface chlorophyll-a concentration (Chl) derived from these satellite observations is used as a proxy for phytoplankton biomass. The connection between seasonal Chl cycles (blooms) and stratification was investigated along with the variability of the mixed-layer depth (MLD) and of the wind stress which is a driver of the MLD. The corresponding data sets are presented in the following section. In Sect. 3, the seasonal and interannual variability of Chl and MLD are presented. The possible role of changes in wind stress intensity in generating the differences in MLD observed between the CZCS and SeaWiFS era is also examined. Results are summarized and discussed in Sect. 4.

2 Data

Ocean color time series are provided by the CZCS (November 1978 – June 1986) and SeaWiFS (September 1997–December 2010) missions. Caution is necessary when comparing data from these two time series separated by a 12-year gap. We accordingly used the reprocessed data set generated by [13], who applied the same algorithms and an adapted calibration to both CZCS and SeaWiFS observations, providing two fully compatible 5-year time series. The years 1979–1983 were selected for the CZCS period and 1998–2002 for the SeaWiFS period. The archive used here is made of monthly composites with an 18-km resolution. It excludes shallow waters (depth < 200 m). Readers are referred to [13] for further understanding of how this reprocessing made the two data sets comparable. For the sake of simplicity, these two time periods will be referred to as the "CZCS period" (or "the 1980s") and the "SeaWiFS period" (or "the 2000s"), though they do not correspond to the full lifetime of these satellite missions.

The database of in situ vertical temperature profiles put together by [14] allowed them to build a global monthly MLD time series starting in 1941. They determined MLD using a temperature criterion, as being the depth where the temperature is 0.2 °C different from the temperature at 10 m to avoid aliasing by the diurnal temperature signal [14]. Here we extracted the North Atlantic (30°–50° N and 40°–0° W) monthly MLD over 1979–1983 and 1998–2002.

Surface wind stress is one forcing of the interannual MLD variability. We used the International Comprehensive Ocean–Atmosphere Data Set (ICOADS), which provides gridded monthly summary products extending from 1800 onward, based on an extensive collection of surface marine data. We used the 1° enhanced product extracted monthly over 1979–1983 and 1998–2002, and derived the wind stress amplitude from its zonal and meridional components [15].

Chl, MLD and wind stress monthly fields were averaged on a 2° by 2° grid and interpolations were carried out to fill empty grid cells. Linear interpolations in time were firstly performed in each cell, and only when the gap is of 1 month. Spatial interpolation along the longitude and latitude axes was subsequently performed only for single isolated empty cells. The MLD and wind stress data sets were barely impacted by this processing because their initial coverage was almost full. At the end, all 2° by 2° cells in the study region were filled with MLD and wind stress data, and 90 % were filled with Chl data.

In order to quantify the interannual variability of MLD and wind stress in the North Atlantic over 1979–1983 and 1998–2002, Empirical Orthogonal Function analyses (EOFs) have been performed on anomaly fields (Sect. 3.3). The anomaly fields were derived by subtracting the average monthly values in each 2° by 2° cell determined over the 10 years, i.e., the 5 years of CZCS plus the 5 years of SeaWiFS.

3 Results

3.1 Basin-Scale Spatial Features and Seasonal Cycles

The distributions of Chl and of the maximum MLD (computed over 1998–2002) are displayed in Fig. 1 for the purpose of illustrating the main features of the study region (30°–50° N; 40°–0° W), which includes the subtropical and subpolar phytoplankton regimes.

At subpolar latitudes the annual phytoplankton biomass cycle is dominated by the spring bloom which occurs in response to increases in mean irradiance of the

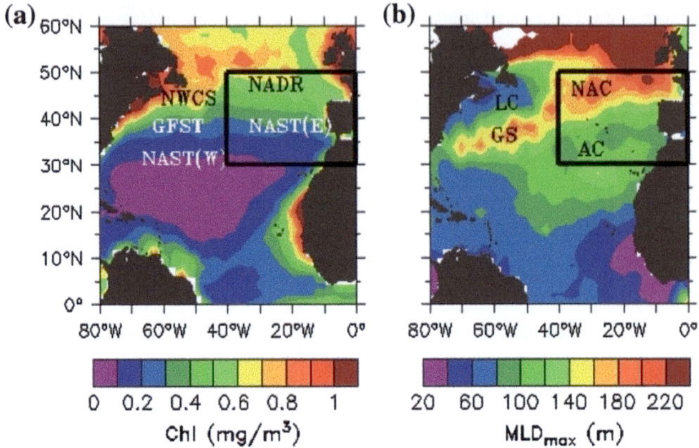

Fig. 1 Average Chl (**a**) and maximum of MLD (**b**), both calculated over the 1998–2002 period in the North Atlantic. The *black* box on both maps delineates the region of our study. The biogeochemical provinces are indicated on **a**: the NWCS, the North Atlantic Drift (NADR), the Gulf Stream (GFST) and the western and eastern North Atlantic Subtropical [respectively NAST(W) and NAST(E)] provinces. The surface currents are indicated on **b**: The Labrador Current (LC), the Gulf Stream (GS), the North Atlantic Current (NAC) and the Azores Current (AC)

mixed layer. At lower latitudes in the subtropics, the biomass peak occurs during winter when mixing by winds and thermal convection replenish the euphotic zone with nutrients. This biomass peak is much reduced in comparison to high-latitude spring blooms [1]. The study region encompasses two of Longhursts' [10] bio-geochemical provinces. In the north, high Chl values appear in the North Atlantic Drift (NADR) province which belongs to the subpolar gyre (Fig. 1a). In the south, low Chl values appear in the eastern part of the North Atlantic Subtropical (NAST) province of the oligotrophic subtropical gyre. Outside the study region, the North Western Continental Shelves (NWCS) province is located northwestward while the Gulf Stream (GFST) province separates the NWCS and NAST provinces. These biogeochemical province boundaries are partly driven by the oceanic dynamics [10]. Therefore, they also appear on the spatial distribution of the winter MLD maximum (MLD_{max}; Fig. 1b). Northwest of 40°–40° W in the NWCS region, MLD_{max} is shallow and the cold Labrador Current flows from the north. The southern boundary of the NWCS region is the Gulf Stream, which flows northeastward and spreads in the northeastern region into the North Atlantic Current, where MLD_{max} is deep ($> \sim 150$ m). Further south, the Azores Current flows eastward then southeastward and MLD_{max} is about 100 m.

The characteristics of the spring and fall blooms are separately analyzed. The fall bloom refers to a Chl peak between September and January, and the spring bloom to a peak between February and August. These two time intervals cover the full year when taken together. The average and standard deviation of the Chl and MLD seasonal cycles over the productive region (from 40°–50° N and 40°–0° W; i.e. the NADR province) are shown in Fig. 2. The Chl seasonal cycle during the

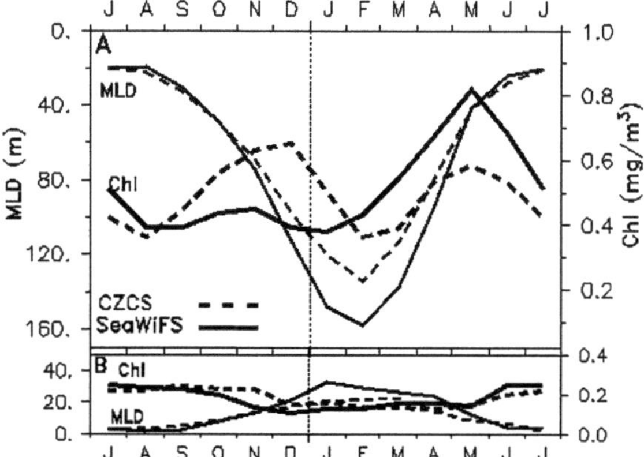

Fig. 2 Time series of means (*top*) and standard deviations (*bottom*) of MLD (*left* axis, *thin lines*) and Chl (*right* axis, *bold lines*). Data are averaged over the Northeastern Atlantic (40°–0° W; 40°–50° N) for the CZCS (*dashed lines*) and SeaWiFS (*solid lines*) era. Time axis is from July to June

period 1979–1983 exhibited a fall bloom (Fig. 2a, 1st peak of the bold dashed line) slightly stronger than the spring bloom (2nd peak, in May). In 1998–2002, the fall bloom was weaker and the spring bloom was dominant (Fig. 2a, bold continuous line). This dominant spring bloom was preceded by a maximum of MLD that was deeper during the SeaWiFS era than it was during the CZCS era (Fig. 2a, respectively dash versus continuous thin line).

3.2 Chl and MLD Interannual Variability in 1979–1983 and 1998–2002

The interannual variability of Chl and MLD is now investigated. The NADR and eastern NAST provinces are located east of 40° W where both Chl and MLD spatial distributions are essentially zonal (Fig. 1). Therefore, Hovmoller plots of Chl and MLD, which show their time variability along latitude in the eastern Atlantic, are simply performed by averaging data along 40°–0° W. They are displayed on Fig. 3.

During the CZCS era (Fig. 3a), the spring bloom started earlier (March–April) in the transition zone (~40° N) than further north in the subpolar gyre (May–June at 50° N). The spring bloom is initiated in the transition zone by the supply of nutrients from winter mixing, while further north in the subpolar gyre, where the ecosystem is likely to be light-limited, the bloom occurs later when the MLD becomes shallower than the critical depth [4, 5]. Conversely, the fall bloom started earlier in the subpolar gyre (in October–November at 50° N), which is about 1–2 months earlier than at 45° N. The fall bloom is initiated in the subpolar gyre when the mixed layer deepens at the end of summer and is progressively refueled with nitrate. Further south, a longer time is necessary for the deepening of the MLD to reach the nutrients that are deeper than northward. It is noteworthy that the fall bloom was of similar amplitude or even higher than the spring bloom.

During the SeaWiFS time period (Fig. 3b), the spring bloom was stronger than in the 1980s, and it extended further south than during the CZCS era, down to

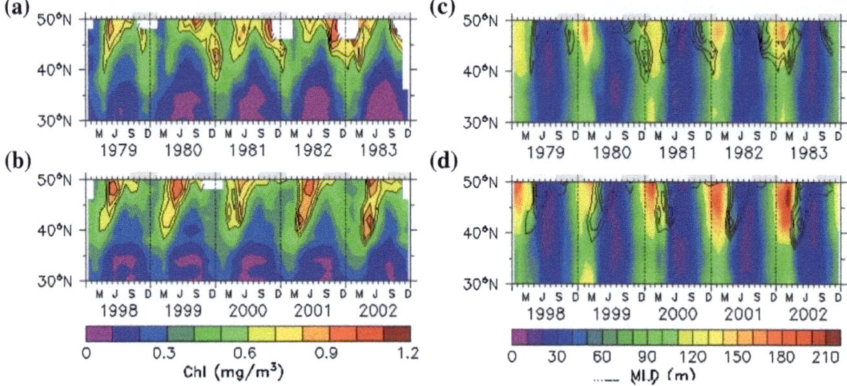

Fig. 3 Latitude versus time plots for the eastern Atlantic (average over 40°–0° W): (**a, b**) Chl and (**c, d**) MLD during the CZCS (**a, c**) and Sea WiFS (**b, d**) era. Chl contours are plotted every 0.1 mg m^{-3} from 0.6 mg m^{-3}. They are reported on the MLD figures. The *grey bars* on the top of the figures correspond to the September-January time period (months of occurrence of the fall bloom)

36°–40° N. The fall bloom was weaker and restricted to the northern boundary. It is also worth noting that the fall bloom in the NADR region was earlier in the 2000s (around September to October; Figure not shown), than it was in the 1980s.

The southward extension of the spring bloom in the 2000s appeared after the occurrence of deeper MLD values (up to 80 m) in winter (Fig. 3c, d). This wintertime increase of MLD in a nutrient limited region likely allowed a more efficient uplift of deep nutrients. In 2001 and 2002, MLD over 36°–50° N was deeper and spring blooms were stronger than in any other year considered here. The weaker fall bloom in the 2000s followed a change in the timing of the MLD deepening at the end of summer, which occurred 1 month later at the beginning of the 2000s than 20 years earlier (September versus August, Figure not shown).

3.3 Wind Forcing and MLD Differences

We investigated the possible role of the wind stress variability on generating the deeper MLD$_{max}$ observed in the 2000s in the eastern Atlantic. For this purpose, we performed EOF analyses on the non seasonal signals of MLD and wind stress (Fig. 4). The first mode of variability of both parameters represents approximately the same amount of their total variance (21 and 26 % respectively for the MLD and wind stress). Their spatial distributions both show high variability in the NADR region (Fig. 4b, c). The MLD and wind stress time functions of the EOFs are driven by the high variability in the northeastern region (Fig. 4a). They are accordingly correlated (r = 0.5). They show similar positive anomalies in the 2000s compared to the 1980s. The positive peaks from 2000 to 2002 both appear on the wind stress and MLD anomalies, suggesting that deeper MLD during the SeaWiFS era were related to stronger winds in the NADR region.

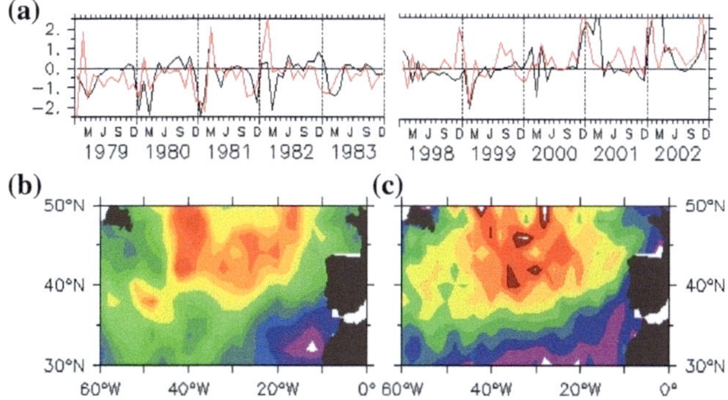

Fig. 4 Time functions and spatial patterns of the first non seasonal mode of the MLD EOFs (*black* curve on **a**, and panel **b**) and wind stress EOFs (*red curve* on **a**, and panel **c**)

4 Summary and Discussion

In the eastern Atlantic (30°–50° N and 40°–0° W), two blooms of similar amplitude occurred in fall and spring at the beginning of the 1980s (CZCS era, 1979–1983). At the beginning of the 2000s (SeaWiFS era, 1998–2002), the spring bloom was stronger and extended further south than in the 1980s. This stronger spring Chl may be related to stronger wintertime winds, which induced deeper MLD, leading to enhanced nutrient uplift (bottom-up process; e.g., [6]). The fall bloom was weaker in the 2000s and only appeared at the northern boundary of the study region. The MLD deepening at the end of summer has been delayed. It occurred 1 month later in 1998–2002 than in 1979–1983. This delay combined with a deeper winter MLD likely increased light limitation and consequently limited a full development of the fall bloom.

The fall bloom is less documented in the literature than the spring bloom. One possible reason is its small amplitude in the 2000s. The way in which time series are often analyzed might also be an explanation, as the usual representation uses plots starting in January, which emphasizes spring blooms, not fall or winter blooms. The weaker fall bloom in the 2000s than in the 1980s is consistent with a 1 month delay in the MLD deepening at the end of summer. This delay we observed in the NADR region has also been reported in the Mid-Atlantic Bight [16]. Although the difference in the timing of the MLD deepening in fall has been put forward here as one possible explanation for the much lower fall bloom in the 2000s, other causes are possible. They include lateral advection [6], mesoscale motions [17] or wind bursts and storms, which all influence the timing of the ML shoaling [7], and can impact the bloom dynamics. One of the most obvious forcing of the differences in the phytoplankton fall bloom would be the frequency of wind bursts in fall and winter, particularly north of 40° N [7]. However the investigation of the frequency of the wind bursts is impeded by the unavailability of high-quality long-term wind records with a sufficiently high temporal resolution.

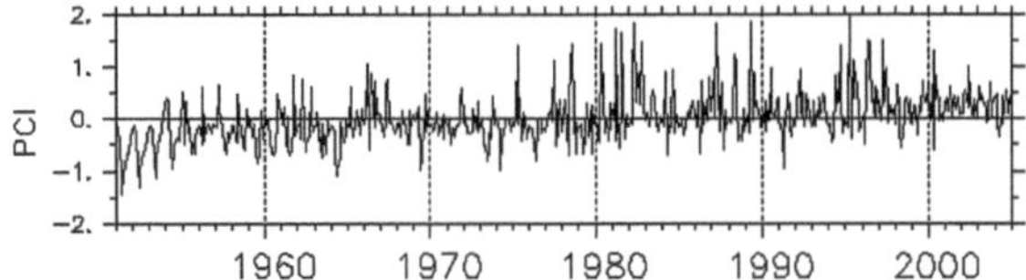

Fig. 5 Time evolution of the anomalies of the Phytoplankton Color Index. Data were spatially averaged over 40°–0° W and 40°–50° N and the anomalies were derived by subtracting the average monthly values over 1951–2004

Sea surface temperature (SST) can be used as an indicator of the ocean stratification, which is one forcing of phytoplankton variability [18]. An inverse relationship between Chl and SST relationship is usually expected in nutrient-limited subtropical regions, because a warming-induced stratification would reduce the upward nutrient supply and then productivity [19, 20]. Stratification would conversely increase productivity in subpolar gyres which are light-limited because of intense vertical mixing [19]. However, a parallel increase of Chl and SST was shown in the North Atlantic from the 1980s to the 2000s [21], and related to a regime shift of the Atlantic Multidecadal Oscillation from a cold to a warm phase in the mid-1980s. The usual scenario would have predicted a shallower MLD and weaker Chl. Our results show that the reported warmer SST in the 2000s than in the 1980s was paralleled to stronger winds and deeper winter MLD_{max}, illustrating that SST can be an ambiguous indicator of stratification. As the present study focuses on two 5-year time periods separated by about 15 years, these differences might also simply reflect interannual variability. However, the correlative deeper MLD and stronger winds that we observed here were also reported over a longer and uninterrupted time series (1960–2004) in this region [22, 23]. These results show that our observations might as well be related to a long-term change. Such a long term trend in the area of our study appears regarding the time evolution of the semi-quantitative measurements of Chl anomalies (Phytoplankton Color Index, PCI) which has been collected by the Continuous Plankton Recorder, [24] on Fig. 5. These results highlight the complexity of the links between SST changes, upper-layer stratification, and biological responses.

Acknowledgments EM was supported by the Changing Earth Science Network initiative funded by the Support To Science Element (STSE) program of the European Space Agency (ESA). This work was supported by the Centre National d'EtudesSpatiales (CNES).

References

1. Yoder JA, McClain CR, Feldman GC, Esaias WE (1993) Annual cycles of phytoplankton chlorophyll concentrations in the global ocean—a satellite view. Glob Biogeochem Cycles 7:181–193
2. Colebrook JM (1979) Continuous plankton records: seasonal cycles of phytoplankton and copepods in the Atlantic Ocean and North Sea. Mar Biol 51(1979):23–32

3. Ducklow HW, Harris RP (1993) Introduction to the JGOFS North Atlantic bloom experiment. Deep Sea Res II 40:1–8

4. Siegel DA, Doney SC, Yoder JA (2002) The North Atlantic spring phytoplankton bloom and Sverdrup's critical depth hypothesis. Science 296:730–733

5. Sverdrup HU (1953) On conditions for the vernal blooming of phytoplankton. Journal du Conseil International pour l'Exploration de la Mer 18:287–295

6. Dutkiewicz S, Follows M, Marshall J, Gregg WW (2001) Interannual variability of phytoplankton abundances in the North Atlantic. Deep Sea Res II 48:2323–2344

7. Lévy M, Lehahn Y, André JM et al (2005) Production regimes in the Northeast Atlantic: a study based on sea-viewing wide field-of-view sensor (SeaWiFS) chlorophyll and ocean general circulation model mixed layer depth. J Geophys Res 110:C07S10. doi:10.1029/2004JC002771

8. Behrenfeld MJ (2010) Abandoning Sverdrup's critical depth hypothesis on phytoplankton blooms. Ecology 91(4):977–989

9. Banse K (1992) Grazing, temporal changes of phytoplankton concentrations, and the microbial loop in the open sea. In: Falkowski PG, Woodhead AD (eds) Primary productivity and biogeochemical cycles in the sea. Plenum Press, New York, pp 409–440

10. Longhurst A (1995) Seasonal cycles of pelagic production and consumption. Prog Oceanog 36:77–167

11. Hovis WA, Clark DK, Anderson F et al (1980) Nimbus-7 coastal zone color scanner: system description and initial imagery. Science (New Series) 210(4465):60–63

12. McClain CR, Feldman G, Hooker S (2004) An overview of the SeaWiFS project and strategies for producing a climate research quality global ocean bio-optical time series. Deep Sea Res II 51:5–42

13. Antoine D, Morel A, Gordon HR, Banzon VF, Evans RH (2005) Bridging ocean color observations of the 1980s and 2000s in search of long-term trends. J Geophys Res 110:C06009. doi:10.1029/2004JC002620

14. De Boyer Montégut C, Madec G, Fischer AS, Lazar A, Iudicone D (2004) Mixed layer depth over the global ocean: an examination of profile data and a profile-based climatology. J Geophys Res 109:C12003. doi:10.1029/2004JC002378

15. Woodruff SD, Worley SJ, Lubker SJ et al (2010) ICOADS Release 2.5: extensions and enhancements to the surface marine meteorological archive. Int J Climatol 31:951–967

16. Schofield O, Chant R, Cahill B et al (2008) The decadal view of the mid-atlantic bight from the COOL room: is our coastal system changing? Oceanography 21(4):108–117

17. McGillicuddy DJ, Robinson AR (1997) Eddy induced nutrient supply and new production in the sargasso sea. Deep Sea Res I 44:1427–1449

18. Behrenfeld MJ, O'Malley RT, Siegel DA et al (2006) Climate-driven trends in contemporary ocean productivity. Nature 444(7120):752–755

19. Doney SC (2006) Oceanography: plankton in a warmer world. Nature 444:695–696. doi:10.1038/444695a

20. Polovina JJ, Howell EA, Abecassis M (2008) Ocean's least productive waters are expanding. Geophys Res Lett 35:L03618. doi:10.1029/2007GL031745

21. Martinez E, Antoine D, D'Ortenzio F, Gentili B (2009) Climate-driven basin-scale decadal oscillations of oceanic phytoplankton. Science 36:1253–1256

22. Yu L (2007) Global variations in oceanic evaporation (1958–2005): the role of the changing wind speed. J Clim 20:5376–5390. doi:10.1175/2007JCLI1714.1

23. Carton JA, Grodsky SA, Liu H (2008) Variability of the oceanic mixed layer, 1960–2004. J Clim 21:1029–1047

24. Reid PC, Matthews JBL, Smith MA (2003) Achievements of the continuous plankton recorder survey and a vision for its future. Prog Oceanogr 58:115–358

INCUSAR—A Method to Retrieve Temporal Averages of 2D Ocean Surface Currents from Synthetic Aperture Radar Doppler Shift

Knut-Frode Dagestad, Morten W. Hansen, Johnny A. Johannessen and Bertrand Chapron

Abstract Within the last decade the Doppler anomaly retrieved from the Synthetic Aperture Radar (SAR) has been demonstrated to be a useful resource for detection of ocean surface currents. As with any Doppler-based method, only the component along the instrument look direction can be retrieved at a time. It is here demonstrated how a time-averaged two-dimensional surface current field can be obtained from corresponding averages of surface current components detected along two non-orthogonal view angles, from respectively ascending and descending passes of the Envisat satellite. By a quantitative assessment of the involved uncertainties, an improved surface current field is obtained by merging the Doppler current with geostrophic current calculated from a climatological Mean Dynamic Topography.

Keywords SAR · Doppler centroid anomaly · Wind field retrieval · Ocean surface currents

1 Introduction

Ocean currents play an important role in the climate system by transporting heat towards the poles, and are also important for marine life by distributing oxygen and nutrients both horizontally and vertically. Monitoring of ocean currents is thus important to quantify and understand such processes, but is also of practical

K.-F. Dagestad (✉) · M. W. Hansen · J. A. Johannessen
Nansen Environmental and Remote Sensing Center, Bergen, Norway
e-mail: Knut-Frode.Dagestad@stormgeo.com

B. Chapron
Laboratoire d'Océanographie Spatiale, Ifremer, Plouzané, France

D. Fernández-Prieto and R. Sabia, *Remote Sensing Advances for Earth System Science*, 59
SpringerBriefs in Earth System Sciences, DOI: 10.1007/978-3-642-32521-2_7,
© The Author(s) 2013

importance for human activities both offshore and in the coastal zone. Satellite remote sensing has become invaluable for monitoring the state of the ocean, but unfortunately ocean surface current is one of the most challenging parameters to retrieve from space. The most widely used method is altimetry, where currents are estimated from the measured slope of the surface, under the assumption of balance between the pressure forces and the Coriolis force (geostrophic assumption). Whereas altimetry provides global coverage, the coarse resolution of the order of 50 km does not resolve the energetic variability of the ocean at mesoscales ($\sim$ 2–30 km). Furthermore, the divergent component of the currents which leads to vertical motion is also not detected as a consequence of the geostrophic assumption. Along track interferometry (ATI) from SAR is a technique with potential to overcome these limitations [8], but this requires a dedicated configuration with two satellites or a split-antenna system, of which the German TerraSAR-X satellites are today the only operational system. In this chapter we consider a method to retrieve the ocean surface current from conventional single-antenna SAR sensors.

2 Retrieval of Line-of-Sight Surface Currents from Envisat ASAR

A Synthetic Aperture Radar detects not only the magnitude of the radar backscatter, but also its phase. From this a Doppler shift can be retrieved, which, in combination with a gross oversampling, is used to provide the fine spatial resolution of the SAR backscatter image. For a space borne instrument such as ASAR onboard the Envisat satellite, a ground pixel size of the order of 10 m can be obtained with such a "synthetic aperture", although the footprint of a single radar beam has a width of the order of 5 km. It has also been demonstrated that the Doppler shift is in itself a useful geophysical product for oceanography [1, 4, 9]. The European Space Agency (ESA) has since 2007 provided a grid of Doppler Centroid (median) values within the Envisat ASAR Wide Swath products. This Doppler Centroid is provided with a pixel spacing of the order of 5 km. A finer resolution could in principle be provided, as a trade-off versus accuracy. However, as shown below, the accuracy is already a critical issue with the given resolution.

The major contribution to the Doppler shift comes from the relative velocity of the satellite and the rotating earth, and can be precisely subtracted to yield the contribution from surface motions relative to the fixed Earth. For Envisat ASAR, some significant instrumental biases distort the result, and corrections are necessary [2]. A Doppler velocity can then be calculated from the corrected Doppler shift with the well known Doppler equation [1, 7]. Within the area over which the Doppler shift is averaged, there will be contributions from the line of sight current component, but also from surface scatterers moving on (as) waves. Averaged over scales of the order of 5 km, it is fair to assume that the Doppler velocity is well approximated by the sum of these two contributions. Although models have been developed

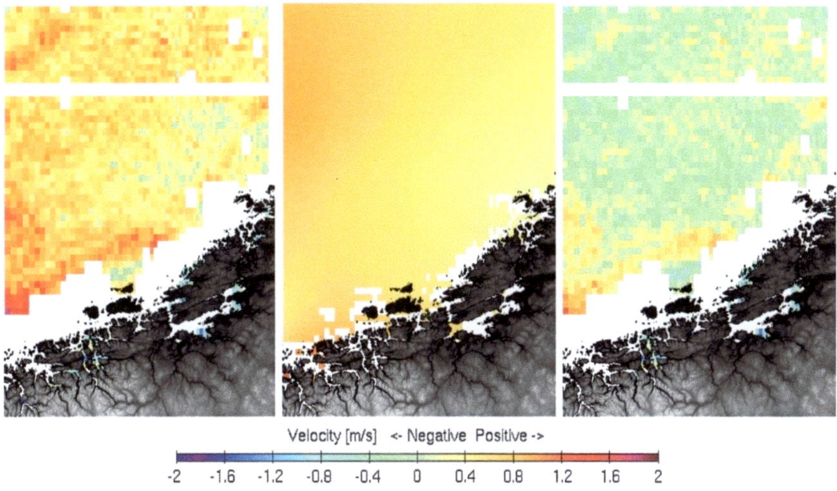

Fig. 1 **a** Range Doppler velocity field from an Envisat ASAR wide swath scene off the west coast of Norway (around Trondheim) on 19 October 2011 at 20:51 UTC. **b** Wind Doppler as calculated with the CDOP function taking a NCEP GFS model wind field as input. **c** The full range Doppler surface velocity minus the wind induced Doppler (**a**, **b**). The white areas are regions where the error corrections have too large uncertainty [2]. By convention, positive values indicate velocity towards *right*, and negative values indicate velocity towards *left*

e.g. [4] it is more convenient and accurate in practice to estimate the wave-state contribution using an empirical relationship between the Doppler velocity and near surface wind speed and direction, CDOP, valid for radar C-band [5].

One example of a Doppler velocity field is shown in Fig. 1a. Figure 1b shows the wave-state contribution, estimated with the CDOP model and the wind field (not shown) from the Global Forecasting System of the National Centers for Environmental Prediction. Subtracting this part from the total Doppler gives an estimate of the component of the surface current along the SAR look direction (Fig. 1c). Despite an uncertainty of the total Doppler velocity as large as 25–40 cm/s, and an uncertainty of the same order introduced through use of the model wind field and the CDOP model [3], one can see a coherent band of positive (rightwards) Doppler velocities along the coast of Norway (Fig. 1c). The velocities (projected to the horizontal) of the order of 50 cm/s agree well with other measurements of the Norwegian Coastal Current, accounting for the fact that the Doppler current includes only the component along the SAR look direction (towards right on Fig. 1).

The uncertainty is in this case comparable to the magnitude of the retrieved surface current component. However, by averaging over time (several acquisitions), this is improved, at the cost of detecting temporally averaged rather than instantaneous currents. As the current direction is unknown at the time of a single SAR acquisition, the Doppler velocities along the ascending and descending directions must be averaged separately. Figure 2 shows such average Doppler

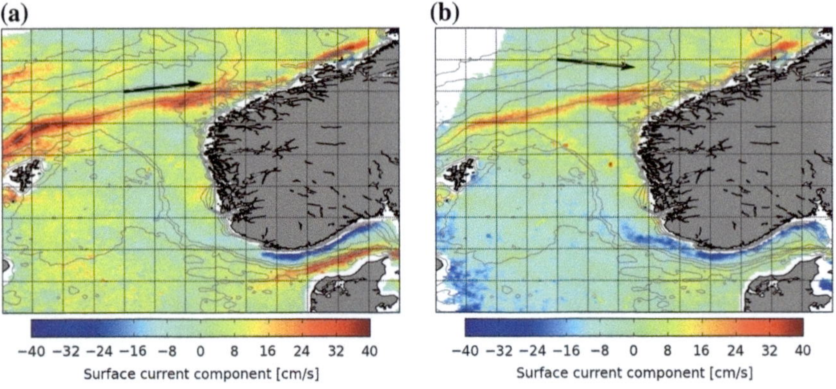

Fig. 2 Averaged **a** ascending and **b** descending ASAR Doppler velocities from [3]. Positive values indicate velocity along the direction of the respective *black arrows*

current components along the ascending and descending look directions from 1,200 scenes acquired between August 2007 and February 2011, from Hansen et al. [3].

The detected current components compare well with both in situ surface drifter measurements, a moored recording current meter, and estimates from combined use of altimetry and gravimetry. Both the Norwegian Atlantic Current, the Skagerrak current and the Norwegian Coastal Current are recognisable in Fig. 2, and most clearly where the current in average has a larger component along the corresponding SAR look direction, which is indicated by the black arrows.

3 Reconstruction of 2D Surface Currents from Line-of-Sight Components

From a small geometrical exercise (see Fig. 3) the orthogonal zonal and meridional velocity components can be calculated:

$$u = \frac{v_a + v_d}{2\cos\alpha}$$

$$v = \frac{v_a - v_d}{2\sin\alpha} \tag{1}$$

where v_a and v_d are the averaged ascending and descending components respectively (as shown in Fig. 2), and α is the angle between the positive directions of the ASAR components and the East direction. For the given latitude α is $13° \pm 1°$.

The reconstructed orthogonal current components are shown in Fig. 4. The meridional (v) component is seen to be much more noisy than the zonal (u)

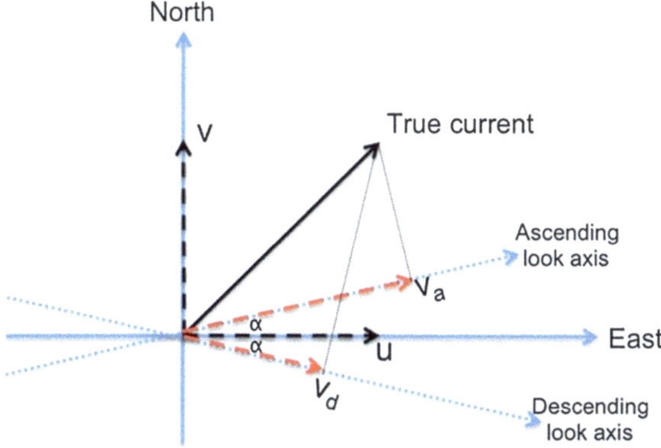

Fig. 3 The *red arrows* illustrate the detected components of the true current vector (*solid black*) for ascending (v_a) and descending (v_d) passes. The *dashed black lines* indicate the orthogonal zonal and meridional components

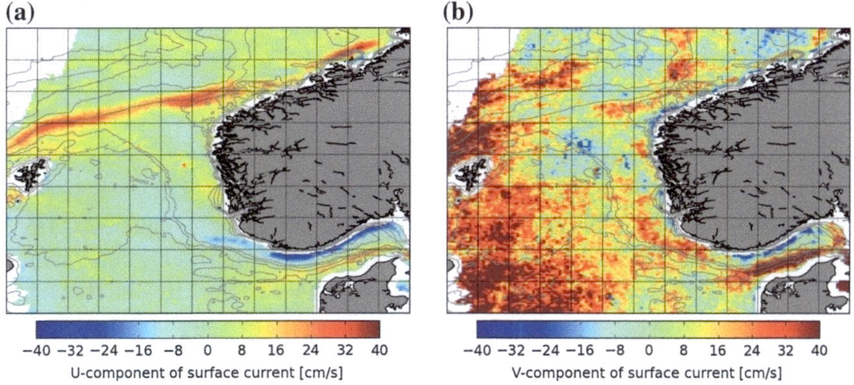

Fig. 4 a zonal and **b** merional components from Envisat ASAR from Eq. (1) and the range directed velocities shown in Fig. 2. The corresponding uncertainties are shown in Fig. 5 c and e

component. This is not surprising since both the ascending and descending look directions are much closer to the zonal (east–west) direction.

This difference of errors can be quantified using the formula for the standard deviation of a general function $f(x_i)$:

$$\sigma_f = \sqrt{\sum_i \left(\frac{\partial f}{\partial x_i}\right)^2 \sigma_i^2} \tag{2}$$

where σ_i is the standard deviation of the variable x_i . Inserting Eq. (1) into Eq. (2) we get the following expressions for the error (standard deviation) of the zonal and meridional current components:

$$\sigma_u = \frac{\sqrt{\sigma_d^2 + \sigma_a^2}}{2\cos\alpha} = 0.73\sigma$$

$$(3)$$

$$\sigma_v = \frac{\sqrt{\sigma_d^2 + \sigma_a^2}}{2\sin\alpha} = 3.14\sigma$$

The latter equality of each equation is valid if the ascending and descending uncertainties are equal ($\sigma_a = \sigma_d = \sigma$). In this case the error of the u component is 27 % smaller than for the ASAR range components, whereas the error of the v component is 314 % larger than the error of the range components. For $\sigma \approx 5$ cm/s as estimated by Hansen et al. [3] we get errors of respectively 3.7 and 15.7 cm/s for the u and v components, in consistence with Fig. 4.

In addition to the variability (noise) in the v-component shown in Fig. 4b, a mean northwards velocity of about 30 cm/s is also seen for most of the British shelf areas. Whereas there are tidal currents with speeds up to and exceeding 1 m/s in this area, such a significant mean current is clearly erroneous, and the values also exceed the uncertainty of the v-component of about 15 cm/s according to Eq. (3) and Fig. 5e. However, in addition to noise, we may also have unknown mean biases Δv_a and Δv_d in the ascending and descending components, which will give a corresponding bias of the v-component of:

$$\Delta v = \frac{\Delta v_a - \Delta v_d}{2\sin\alpha} \qquad (4)$$

Thus biases of equal magnitude but opposite sign will lead to a bias in the v-component which is 5 times larger than that of the range components. Hence moderate biases of e.g. +6 and −6 cm/s for the range components could explain the large bias of 30 cm/s of the v-component observed in Fig. 4b. Remaining instrumental biases [2] could be one possible explanation for this observed artifact. Additionally, the sun-synchronous orbit of Envisat may also lead to a significant sampling aliasing of tidal currents which are in this region primarily oscillating with the lunar semi-diurnal period of 12.4 h.

The combined mean 2D surface current field is shown in Fig. 5a. Except for the British shelf, the current vectors are reasonable both in magnitude and direction. The main features compare well with the geostrophic current shown in Fig. 5b, which is calculated from a climatological Mean Dynamic Topography based on combination of GRACE data, altimetry and in situ measurements [6].

As obtained (Fig. 5 c−f) the u-component of the Doppler current is more accurate than the corresponding geostrophic component. For the v-component, this is reversed, except for some areas close to the coast, where satellite altimetry has limitations. One indication of the latter limitation is the absence of the coastal current along the coast of Southwest Norway, which is visible on the SAR Doppler map, despite the south-north orientation of the current which is unfavourable for the Doppler detection.

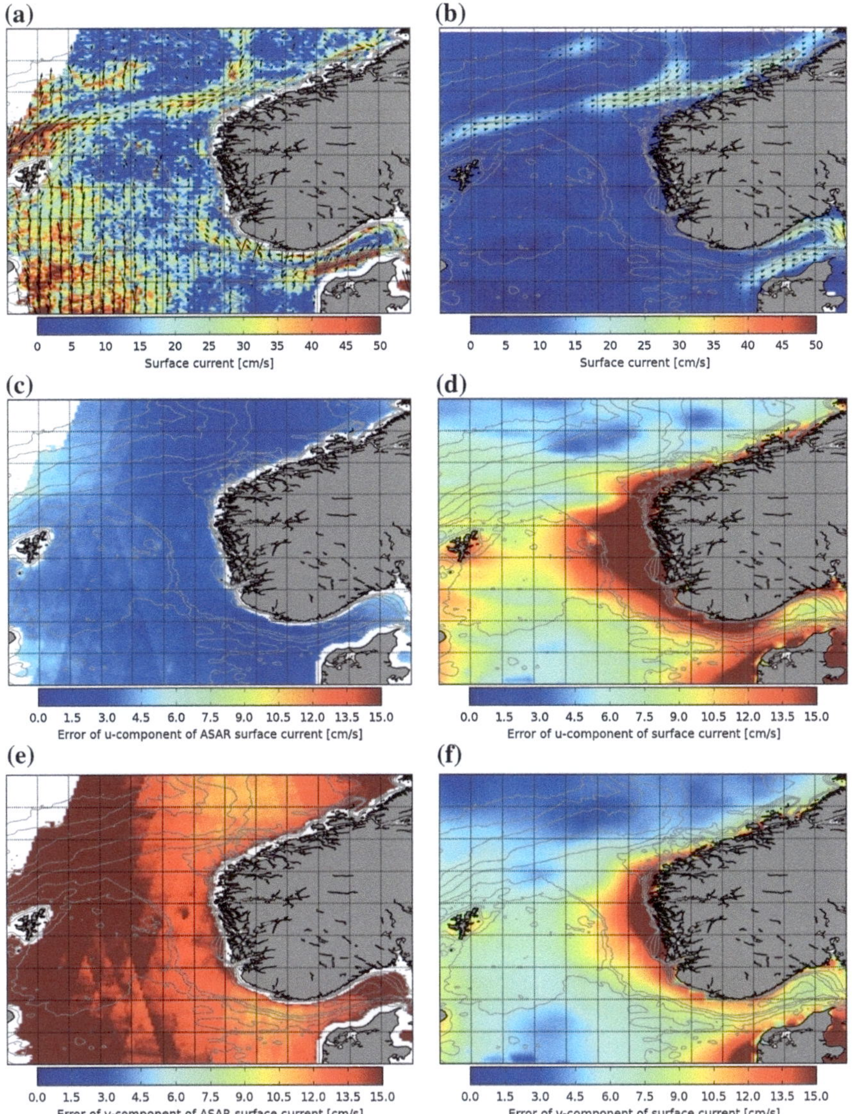

Fig. 5 a Surface currents from Envisat ASAR Doppler shift, by combining the components shown in Fig. 4. **b** Geostrophic current from [6]. **c** and **d** show the errors (standard deviations) of the u-components of the ASAR current (**a**) and the geostrophic current (**b**), and **e** and **f** show errors of the corresponding v-components

To take maximum benefit from the strengths of these two different sources of information, they can be merged. Following a Bayesian approach, we can calculate an optimum meridional velocity v' by

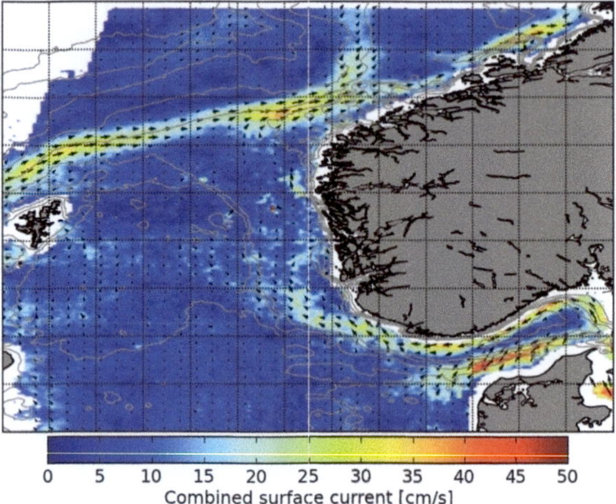

Fig. 6 Surface current calculated as a weighted combination of the ASAR Doppler current and the geostrophic current from a mean dynamic topography of Rio et al. [6]

$$v' = \frac{v\sigma_{gv}^2 + v_g\sigma_v^2}{\sigma_{gv}^2 + \sigma_v^2},\tag{5}$$

where σ_{vg} is the error of the v-component (v_g) of the geostrophic current. A corresponding equation can be found for a combined zonal component u'. The resulting merged ASAR and geostrophic current is shown in Fig. 6. The erroneous northwards current on the British shelf is removed, as the v-component of the Doppler current with high uncertainty obtained little weight here. The signatures of the northwards flowing Norwegian coastal current are however preserved, the geostrophic current having lower accuracy at this location. Overall, this final result shows a more consistent and coherent mean current system connecting the Norwegian Atlantic Current with the Norwegian Coastal Current and the Skagerrak current.

For the calculation of the Doppler current, we corrected for the wave state by using the wind speed relative to a non-moving ocean surface. Given now a first guess current field, this procedure could be repeated by using a wind speed relative to the first guess current, presumably giving further improvements.

4 Conclusions

It is demonstrated how temporally averaged ocean surface current vectors can be retrieved from corresponding non-orthogonal SAR range Doppler components. The uncertainty is shown to be significantly higher for the retrieved meridional (v) component due to unfavourable satellite view configuration. A weighted combination with a climatological geostrophic current improves the accuracy, while preserving the finer spatial details (~ 10 km) provided by the SAR measurements.

Whereas the SAR measurements are here averaged over several years, improved specifications for future SAR systems (e.g. Sentinel-1) may yield the proposed method applicable to resolve seasonal and monthly variations of even moderately weak ocean currents.

References

1. Chapron B, Collard F, Ardhuin F (2005) Direct measurements of ocean surface velocity from space: interpretation and validation. J Geophys Res 110:C07008. doi:10.1029/2004JC002809
2. Hansen MW, Collard F, Dagestad K-F, Johannessen JA, Fabry P, Chapron B (2011a) Retrieval of sea surface range velocities from Envisat ASAR Doppler centroid measurements. IEEE Trans Geo R Sens 49(10):3582–3592. doi: 10.1109/TGRS.2011.2153864
3. Hansen MW, Johannessen JA, Dagestad K-F, Collard F, Chapron B (2011b) Monitoring the surface inflow of atlantic water to the norwegian sea using Envisat ASAR. J Geophys Res. doi:10.1029/2011JC007375
4. Johannessen JA, Chapron B, Collard F, Kudryavtsev V, Mouche A, Akimov D, Dagestad K-F (2008) Direct ocean surface velocity measurements from space: improved quantitative interpretation of Envisat ASAR observations. Geophys Res Lett 35:L22608. doi:10.1029/2008GL035709
5. Mouche A, Collard F, Chapron B, Dagestad K-F, Guitton G, Johannessen J, Kerbaol V, Hansen MW (2012) On the use of Doppler shift for sea surface wind retrieval from SAR. IEEE Trans Geo R Sens 50(7):2901–2909
6. Rio MH, Guinehut S, Larnicol G (2011) The new CNES-CLS09 global mean dynamic topography computed from the combination of GRACE data, altimetry and in situ measurements. J Geophys Res-Oceans. doi:10.1029/2010JC006505
7. Romeiser R, Thompson DR (2000) Numerical study on the along track interferometric radar imaging mechanism of oceanic surface currents. IEEE Trans Geosci Remote Sens 38:446–458
8. Romeiser R, Thompson D (2001) Study on concepts for radar interferometry from satellites for ocean and land applications, Final report of the KoRIOLiS project. http://www.ifm.zmaw.de/projectsites/bmbf-finanziert/beginn-2001/koriolis/
9. Rouault MJ, Mouche A, Collard F, Johannessen JA, Chapron B (2010) Mapping the Agulhas current from space: an assessment of ASAR surface current velocities. J Geophys Res 115:C10026. doi:10.1029/2009JC006050

OC-Flux—Open Ocean Air-Sea CO$_2$ Fluxes from Envisat in Support of Global Carbon Cycle Monitoring

Jamie D. Shutler

Abstract Increasing levels of atmospheric carbon dioxide gas (CO$_2$) from anthropogenic sources are of growing concern due to their impact on the global climate system. Research has shown that there is a strong link between increasing atmospheric CO$_2$ concentrations and the warming climate. The global oceans are considered the only true net sink of atmospheric CO$_2$. Understanding the exchange of CO$_2$ between the ocean and the atmosphere is clearly of importance for monitoring the global carbon cycle and for climate modelling. However, there remain large uncertainties in the current parameterisations of air-sea gas interactions, which can have profound effects on the resulting global estimates of CO$_2$ uptake. The OC-flux project was funded to exploit coincident Envisat data to investigate a number of these uncertainties.

Keywords Air-sea fluxes · Gas transfer velocity · Carbon cycle

1 Introduction

Increasing levels of atmospheric CO$_2$, caused by the burning of fossil fuels and biomass, are of growing concern due to their impact on the global climate system. Currently, it is thought that the ocean system annually absorbs up to $\sim$25 % of anthropogenic CO$_2$ [24]. But it is not clear whether this ocean carbon 'sink' is increasing or decreasing, and considerable temporal and spatial inter-annual variation appears to occur e.g. [32]. It is therefore crucial to identify the associated uncertainties in these estimates [14]. Space observations (or Earth observation)

J. D. Shutler (✉)
Plymouth Marine Laboratory, Plymouth PL1 3DH, UK
e-mail: jams@pml.ac.uk

D. Fernández-Prieto and R. Sabia, *Remote Sensing Advances for Earth System Science*, SpringerBriefs in Earth System Sciences, DOI: 10.1007/978-3-642-32521-2_8,
© The Author(s) 2013

have an important role to play in this process through providing quasi-synoptic, reproducible and well-calibrated measurements for investigating processes on global scales.

The flux of CO_2 between the atmosphere and the ocean (air-sea) is controlled by wind speed, sea state, sea surface temperature and surface processes including any biological activity. The air-sea flux of CO_2 can be determined from the gas transfer velocity, k, using:

$$F = k(\alpha_W pCO_{2W} - \alpha_S pCO_{2A}) \tag{1}$$

where α is the solubility of the gas in water at depth (α_W) and at the sea skin (α_S) and pCO_2 is the partial pressure of CO_2 in the water at depth (pCO_{2W}) and at the surface (pCO_{2A}). A range of wind speed and sea state dependent parameterisations for k exist e.g. [22]. The solubility (α) of the gas (CO_2) is a known function of sea surface temperature (SST) and salinity.

2 Global Scale Accuracy Assessment

Earth observation (EO) in conjunction with additional datasets allows the estimation and monitoring of air-sea fluxes on a global scale. However, the global accuracy of such methods has yet to be determined. There are currently several approaches for estimating k from EO data. The majority of these approaches exploit the dependence of k on wind speed and the predictions of k vary significantly among these relationships, especially at higher wind speeds, as shown in Fig. 1 (k_{660} is the k for a Schmidt number of 660 which corresponds to CO_2 in sea water at 20 °C). The lack of any global accuracy assessment situation is in part due to the difficulty in collecting suitable ground truth in situ data. However, one of the major hurdles is the lack of temporally and spatially coincident EO SST and wind/wave data, which are required to accurately assess the global performance of the various approaches. However, the European environmental monitoring satellite, Envisat, which carries the Radar Altimeter 2 (RA2), the Advanced Along Track Scanning Radiometer (AATSR) and the Medium Resolution Imaging Spectrometer (MERIS) is capable of providing spatially and temporally coincident EO data for a range of environmental parameters which can be directly compared with coincident in situ data.

3 Surface Biology and Temperature Gradients

The accumulation of biological surfactant at the water surface introduces surface-tension gradients, which can modify both momentum and mass transports in close proximity to the surface [28], effectively attenuating surface roughness [6].

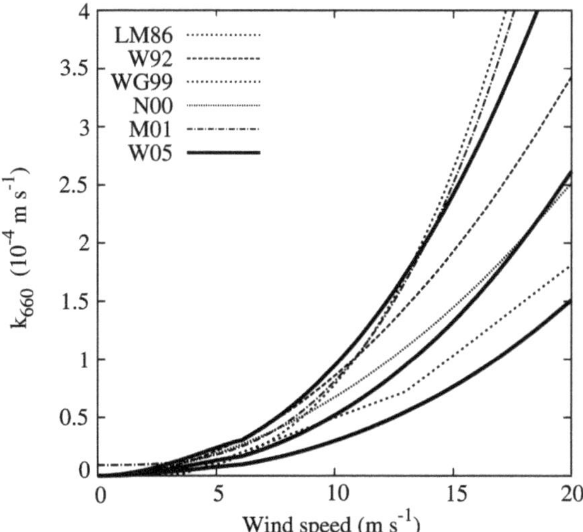

Fig. 1 Gas transfer velocity (k) versus wind speed for a range of different parameterisations for a Schmidt number of 660. [15] (LM86), [29] (W92), [30] (WG99), [22] (N00), [19] (M01), and a family of curves for [33] (W05), which is the basis for the approach of [5]

A number of studies have shown experimental evidence that the presence of surfactants can reduce gas transport across the water surface e.g. [25]. The major cause of such oceanic surfactants are marine phytoplankton and a range of phytoplankton species are able to produce such surface-active materials e.g. [7]. This suggests that primary productivity, which can be estimated from EO data, can be used a proxy for indicating the presence of biological slicks [28].

Temperature gradients can exist at the air-sea interface due to molecular transfer processes and radiative heating, which can impact on the air-sea CO_2 fluxes. Therefore the temperature at the precise water interface (SST_{int}) is the interface with atmosphere and so governs the air-sea exchange process. However, while SST_{int} theoretically exists it is physically impossible to measure. Instead, the skin temperature, SST_{skin}, defined as the temperature within the conductive diffusion-dominated sub-layer is considered to be a good approximation of SST_{int}. Increasingly, in situ studies are highlighting the need to use SST_{skin} to derive air-sea CO_2 fluxes, allowing the impact of temperature gradients to be included in the analysis [20, 31]. Similarly, one-dimensional model studies have highlighted the impact of temperature gradients on the calculated flux e.g. [11]. Collectively, these localised studies have shown that the existence of surface temperature gradients can change the magnitude of the flux by 30–40 % [11, 20]. However, most global air-sea CO_2 flux analyses use sub-skin estimates of SST (e.g. from microwave EO sensors), or model and EO derived depth estimates of SST (e.g. from near-infrared EO sensors) calibrated to sub-surface or SST_{fnd} temperatures [2]. The impact of using skin temperature datasets for representing the surface conditions when deriving global CO_2 fluxes (in contrast to using foundation temperature, SST_{fnd}) could be very large. For instance there are number of regions in the world where the SST_{skin} is expected to deviate considerably from the SST at depth. These regions include the equatorial Pacific and the central Atlantic oceans [4]. Earth observation (EO) is able to provide accurate and precise SST_{skin} data (e.g. from near-infrared EO sensors that are calibrated to determine skin temperatures [2]) and these can be used to represent the water temperature at the air-sea interface.

OC-flux was funded to use coincident data from the European Space Agency's (ESA) Environmental monitoring satellite, Envisat, to investigate uncertainties in current estimates of air-sea fluxes of CO_2. This work exploited advances in methods [9, 13, 26, 27] and expanded climatology datasets [27] to evaluate (i) the accuracy of EO to determine k, (ii) the impact of temperature gradients on these fluxes and (iii) the importance of accounting for biological activity when estimating these fluxes.

4 Methodology

Spatially and temporally coincident sea skin temperature (SST_{skin}), wind speed (U_{10}), significant wave height (H_S), backscatter coefficient (σ_0) and spectral information on surface biology were provided by the Envisat RA2, AATSR and MERIS data. Climatological values including the partial pressure of CO_2 in water at depth (pCO_{2W}) in air (pCO_{2A}) and salinity were provided by the [27] climatology. Global daily air-pressure and air temperature fields were provided by the European Centre for Medium-range Weather Forecasting (ECMWF) operational dataset (N80 Gaussian gridded analysis on surface levels; ERA-40 format, 6 hourly fields). Daily SST_{fnd} and sea ice data were provided by the Operational Sea Surface Temperature and Sea Ice Analysis (OSTIA) system [3]. All data were re-projected to a common $1^\circ \times 1^\circ$ grid. To achieve this the Takahashi data were linearly interpolated and the ECMWF data were re-binned. While Eq. 1 can be used to calculate the air-sea flux of CO_2, a more accurate calculation of the flux should include a small temperature correction which results from the temperature dependent carbonate reaction [8]. Therefore, the flux was computed using the above datasets following:

$$F = k(\alpha_W(1 + 0.015\Delta T)pCO_{2W} - \alpha_S pCO_{2A}) \tag{2}$$

which is Eq. 1 with a temperature correction for the carbonate reaction.

The climatology pCO_{2W} and pCO_{2A} data were pressure and temperature corrected following [12], α_W was determined using SST_{fnd}, whereas α_S was determined using SST_{skin} and $\Delta T = SST_{fnd}-SST_{skin}$. The net global fluxes were calculated by first generating daily mean flux composites (in Pg C day^{-1}) for each month. These daily mean composites were then multiplied by the number of days in the respective month (to produce values in Pg C mon^{-1}). Each $1^\circ \times 1^\circ$ flux value was then multiplied by its respective areal coverage (modeling the Earth as a sphere) and all area-weighted values were then summed to determine the net monthly flux. All monthly calculations were normalised with respect to monthly mean ice cover [27]. Summing the resultant monthly net fluxes across each year

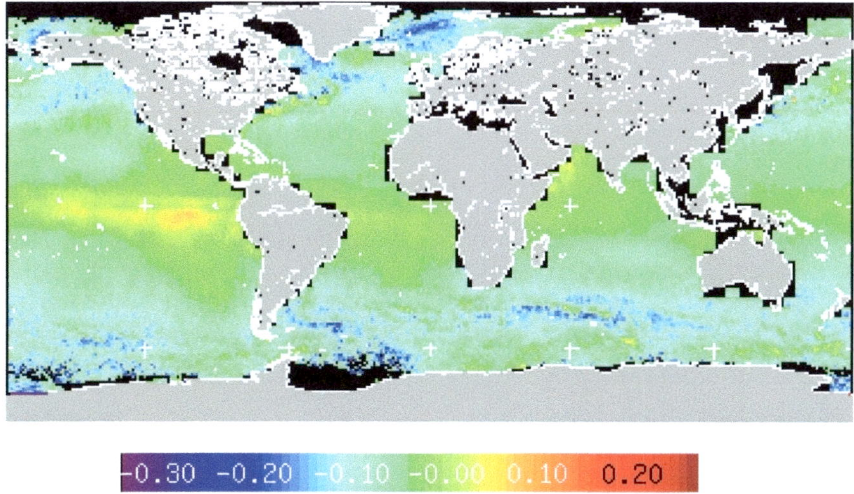

Fig. 2 Daily mean CO_2 flux (g C m^{-2} day^{-1}) for N00 k parameterisations. The flux is calculated assuming satellite data are representative of the whole day and then averaged the whole year. Land is in *grey* and *black* represents regions of no data

produces the net annual flux (in Pg C yr^{-1}). Figure 2 shows the daily mean air-sea CO_2 flux for the N00 parameterisation of k.

5 Evaluating Uncertainties in EO Methods Using In Situ Data

A large in situ k dataset was collated [10, 16–18, 21, 34, 35] and compared with their equivalent EO data. For each in situ data point the closest valid pixel was extracted from the EO datasets. Due to the EO data consisting of binned data, near coincident single pixel values were extracted. To maximize the number of potential matchups the EO data were retained for comparison if i) the Euclidean distance between the in situ sample and EO grid point was $\leq 1°$ (i.e. nearest neighbour, ii) the EO pass and in situ data were collected within $\pm$ 6 h of each other and iii) data existed for SST and U_{10} in both the in situ data and EO data. A number of different statistical quantities were then used to assess the differences between the in situ and EO data including the standard error (S) and root mean squared error (RMSE). Figure 3 shows the scatter plots for in situ k versus EO derived k for two different parameterisations. Also plotted (as error bars) are the associated uncertainties from the EO and in situ data. The EO data uncertainties have been propagated through each k parameterisation calculation using the actual published estimates of uncertainties [1, 23]. While the Fig. 3a shows a slope close

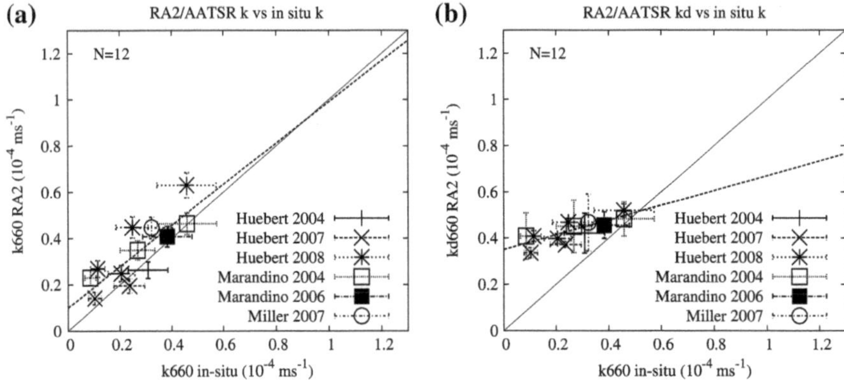

Fig. 3 Envisat derived k versus in situ k for **a** N00 with S = 0.20, RMSE = 0.16, R^2 = 0.66, slope = 0.89, intercept = 0.10, N = 12 and **b** FW07 with S = 0.08, RMSE = 0.25, R^2 = 0.64, slope = 0.32, intercept = 0.35, N = 12

to the 1:1 line for N00, it is important to note that the standard error in Fig. 3b is much lower, suggesting that re-parameterising FW07 (to correct the poor slope response of 0.32) is likely to provide a more superior parameterisation of k (than using N00). It is noted that a much larger dataset is really required to fully evaluate these approaches. Due to the low number of in situ-EO matchups these results only provide an indication of the performance of the different methods over a limited range of environmental conditions.

6 Impact of Vertical Temperature Gradients

In order to isolate the impact of near surface vertical temperature gradients on the global air-sea flux of CO_2, two scenarios are considered i) fluxes generated using daily SST_{skin} and SST_{fnd} data (F_{skin}) and ii) fluxes generated using daily SST_{fnd} data (F_{fnd}). Using the first scenario as the reference dataset, the percentage difference (termed error) in the second scenario was investigated. The error gives an indication of the impact on the fluxes of neglecting vertical temperature gradients. The net global flux for 2008–2009 was determined using F_{skin} to produce a mean net sink of 1.19–2.82 Pg C yr^{-1}, dependent upon the gas transfer velocity parameterisation used. When F_{fnd} is used this sink reduces to 1.06–2.52 Pg C yr^{-1}. Figure 4 shows the mean F_{fnd}–F_{skin} differences for the N00 parameterisations of k. For instance this highlights that if F_{fnd} is used to estimate local fluxes in the Southern Ocean (i.e vertical temperature gradients are ignored) then the flux is likely to be in error by more than 20 %.

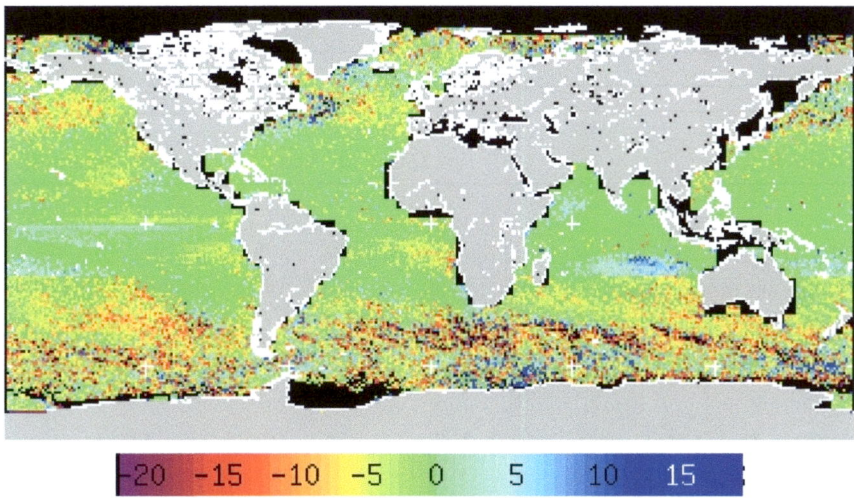

Fig. 4 Percentage differences between the air-sea CO_2 fluxes (F_{fnd}–F_{skin}) calculated using the N00 parameterisation of k. Land is in *grey*; *black* represents regions of no data

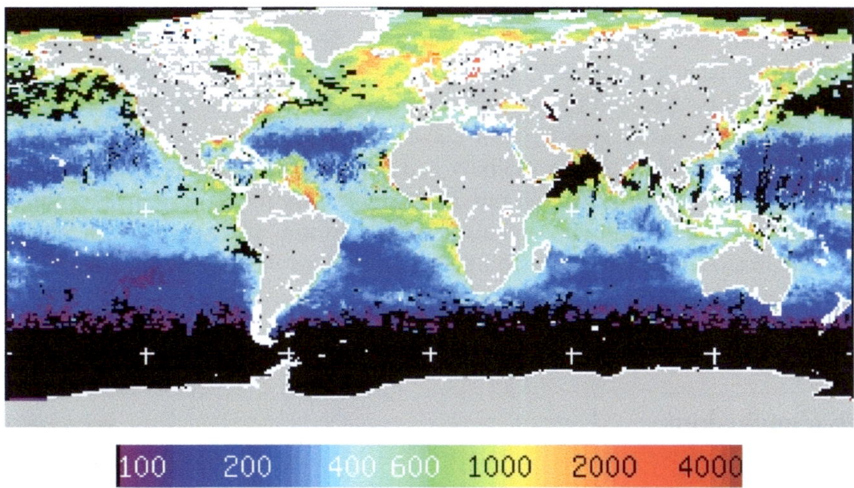

Fig. 5 Monthly mean primary production estimates from coincident AATSR and MERIS data for July 2008 in mg C m^2 day^{-1} using [26]

7 Importance of Surface Biology

Figure 5 shows daily mean net primary production estimated using the coincident Envisat data. These daily estimates were used to quantify the percentage of the net flux which occurs in regions of high biological activity as defined by [28]. Table 1 shows that despite the relatively small areas of open ocean that these biological

Table 1 Global net fluxes (Ψ) and net areal coverage (Φ) and the percentage due to slick (high biology) regions

Parameter		Complete region	% due to slicks
Global Ψ	N00	-1.10 Pg C yr^{-1}	9 (0.10)
	FW07	-1.88 Pg C yr^{-1}	10 (0.18)
Global Φ		2.88×10^{15} km^2	6 (0.18×10^{15})

slicks cover (~ 6 % of the global open ocean) they can have a measurable impact on the net flux (i.e these regions account for ~ 10 % of the global net flux). This impact varies temporally (not shown).

8 The Potential of Sentinel 3 Sensors for Studying Air-Sea Fluxes of CO_2

After 2013 a series of 'Sentinel' platforms are planned to provide operational monitoring of the Earth's environment. The Sentinels are part of the joint European Commission (EC) and European Space Agency (ESA) Global Monitoring for Environment and Security (GMES) initiative. Sentinel 3, the first of which is planned for launch in late 2013, will carry the Synthetic Aperture Radar altimeter (SRAL) and the Sea and Land Surface Temperature Radiometer (SLSTR). These instruments are based upon Cryosat-2 and AATSR concepts and technology, and will be capable of collecting spatially and temporally coincident data. Therefore the Sentinel 3 platforms should provide suitable data for the continuation of the work detailed here. Multiple versions of these platforms are expected with Sentinel 3a and 3b already planned, providing the potential for the long term monitoring of near-real time air-sea CO_2 fluxes.

9 Conclusions

The OC-flux project has exploited synergy between the sensors onboard Envisat to study the atmosphere–ocean flux of CO_2 in the global oceans. This work has resulted in a number of key findings.

It is clear that a larger number of in situ measurements of the gas transfer velocity and air-sea CO_2 fluxes are required to allow Earth observation methods for monitoring air-sea CO_2 fluxes to be fully evaluated. Routine collection of such in situ data is really required if Earth observation is to be fully exploited for monitoring air-sea gas exchange. However, the initial Earth observation and in situ comparison results presented here support the use of sea state based parameterisations in place of purely wind based methods.

Neglecting near surface temperature gradients when deriving global air-sea fluxes of CO_2 can result in global errors of up to 11 % in the annual net fluxes. Locally these errors can be >20 %. These results highlight a large source of error in foundation or sub-surface temperature driven estimates of global net air-sea CO_2 flux that ignore near-surface temperature gradients.

Globally in the open oceans regions of biological slicks account for ~ 10 % of the net flux of CO_2 (sink). Consequently, global flux studies that use gas transfer velocity parameterisations that do not account for the dampening of the water surface due to biological slicks are likely to overestimate the net sink.

Acknowledgments The author would like to acknowledge the helpful discussions from Dr David Woolf, Dr Craig Donlon, Dr Nick Hardman-Mountford, Dr Phil Nightingale, Dr Lonneke Goddijn-Murphy, Dr Diego Fernandez and Dr Steffen Dransfeld. Some of this work was undertaken during a visit (funded by a UK NCEO grant) to the ESA European Space Technology and Research Centre (ESTEC). All RA2 BRAT processing used the ESA Grid Processing On Demand (G-POD). Data post processing was achieved using resources provided by the UK NERC Earth observation Data Acquisition and Analysis Service (NEODAAS). This work was partly funded by the UK NERC marine research programme Oceans 2025.

References

1. Cotton PD, Carter DJT, Challenor PE (2003) Geophysical validation and cross calibration of Envisat RA2 wind wave products at satellite observing systems and Southampton oceanography centre. Envisat RA2 and MWR CCVT. Southampton Oceanography Centre, Southampton
2. Donlon CJ, Casey KS, Robinson IS, Gentemann CL, Reynolds RW, Barton I, Arino O, Stark J, Rayner N, LeBorgne P, Poulter D, Vazquez-Cuervo J, Armstrong E, Beggs H, Jones DL, Minnett PJ, Merchant CJ, Evans R (2009) The GODAE high resolution sea surface temperature pilot project (GHRSST-PP). Oceanography 22(3):34–45
3. Donlon CJ, Martin M, Stark JD, Roberts-Jones J, Fiedler E, Wimmer W (2011) The operational sea surface temperature and sea ice analysis (OSTIA). Remote sensing of environment special issue on (A)ATSR. doi:10.1016/j.rse.2010.10.017
4. Donlon CJ, Minnett PJ, Gentemann C, Nightingale TJ, Barton IJ, Ward B, Murray MJ (2002) Toward improved validation of satellite sea surface skin temperature measurements for climate research. J Clim 15:353–369
5. Fangohr S, Woolf DK (2007) Application of new parameterizations of gas transfer velocity and their impact on regional and global CO_2 budgets. J Mar Syst 66:195–203
6. Frew NM, Bock EJ, Schimpf W, Hara T, Haußecker H, Edson JB, McGillis WR, Nelson RK, McKenna SP, Uz BM, Jähne B (2004) Air-sea gas transfer: its dependence on wind stress, small-scale roughness, and surface films. J Geophys Res 109(C08S17). doi:10.1029/2003JC002131
7. Frew NM, Goldman JC, Dennett MR, Johnson AS (1990) Impact of phytoplankton-generated surfactants on air-sea gas exchange. J Geophys Res 95(C3):3337–3352
8. Hare JE, Fairall CW, McGillis WR, Edson JB, Ward B, Wanninkhof R (2004) Evaluation of the national oceanic and atmospheric administration/coupled-ocean atmospheric response experiment (NOAA/COARE) air-sea gas transfer parameterisation using GasEx data. J Geophys Res 109(C08S11). doi:10.1029/2003JC001831
9. Ho DT, Wanninkhof R, Schlosser P, Ullman DS, Hebert D, Sullivan KF (2011) Towards a universal relationship between wind speed and gas exchange: gas transfer velocities

measured with 3He/SF6 during the Southern Ocean gas exchange experiment. J Geophys Res 116(C00F04). doi:10.1029/2010JC006854

10. Huebert BJ, Blomquist BW, Yang M, Archer SD, Nightingale P, Yelland M, Stephens J, Pascal R, Moat B (2010) Linearity of DMS transfer coefficient with both friction velocity and wind speed in the moderate wind speed range. Geophys Res Lett 37(L01605). doi:10.1029/2009GL041203

11. Jeffery CD, Robinson IS, Woolf DK, Donlon CJ (2008) The response to phase-dependent wind stress and cloud fraction of the diurnal cycle of SST and air-sea CO_2 exchange. Ocean Model 23:32–48

12. Kettle H, Merchant CJ (2005) Systematic errors in global air-sea CO_2 flux caused by temporal averaging of sea-level pressure. Atmos Chem Phys 5:1459–1466

13. Kettle H, Merchant CJ, Jeffery CD, Filipiak MJ, Gentemann CL (2009) The impact of diurnal variability in sea surface temperature on the central Atlantic air-sea CO_2 flux. Atmos Chem Phys 9:529–541

14. Le Quere C, Raupach MR, Canadell JG, al. Me (2009) Trends in the sources and sinks of carbon dioxide. Nat Geosci 2(12):831–836

15. Liss PL, Merlivat L (1986) Air-sea gas exchange rates: introduction and synthesis. The role of air-sea gas exchange in geochemical cycling. Reidel, Washington

16. Marandino CA, De Bruyn WJ, Miller SD, Saltzman ES (2007) Eddy correlation measurements of the air/sea flux of dimethylsulfide over the North Pacific Ocean. J Geophys Res 112(D03301). doi:10.1029/2006JD007293

17. Marandino CA, De Bruyn WJ, Miller SD, Saltzman ES (2008) DMS air/sea flux and gas transfer coefficients form the North Atlantic summertime coccolithophore bloom. Geophys Res Lett 35(L23812). doi:10.1029/2008GL036370

18. Marandino CA, De Bruyn WJ, Miller SD, Saltzman ES (2009) Open ocean DMS air/sea fluxes over the eastern South Pacific Ocean. Atmos Chem Phys 9:345–356

19. McGillis WR, Edson JB, Hare JE, Fairall CW (2001) Direct covariance air-sea CO_2 fluxes. J Geophys Res 106(C8):16729–16745

20. McGillis WR, Edson JB, Zappa CJ, Ware JD, McKenna SP, Terray EA, Hare JE, Fairall CW, Drennan W, Donelan M, DeGrandpre MD, Wanninkhof R, Feely RA (2004) Air-sea CO_2 exchange in the equatorial Pacific. J Geophys Res 109 (C08S02). doi:10.1029/2003JC002256

21. Miller SD, Marandino CA, De Bruyn WJ, Saltzman ES (2009) Air-sea gas exchange of CO_2 and DMS in the North Atlantic by eddy covariance. Geophys Res Lett 36(L15816). doi:10.1029/2009GL038907

22. Nightingale P, Malin G, Law C, Watson AJ, Liss PL, Liddicoat M, Boutin J, Goddard RU (2000) In situ evaluation of air-sea gas exchange parameterizations using novel conservative and volatile tracers. Global Biogeochem Cycles 14:373–387

23. O'Carroll AG, Eyre JR, Saunders RW (2008) Three-way error analysis between AATSR, AMSR-E and in situ sea surface temperature observations. J Atmos Ocean Technol 25:1197–1207

24. Sabine CL, Feely RA, Gruber N, Key RM, Lee K, Bullister JL, Wanninkhof R, Wong CS, Wallace DWR, Tilbrook B, Millero FJ, Peng TH, Hozyr A, Ono T, Rios AF (2004) The oceanic sink for anthropogenic CO_2. Science 305:367–371

25. Salter ME, Upstill-Goddard RC, Nightingale P, Archer SD, Blomquist BW, Ho DT, Huebert BJ, Schlosser P, Yang M (2011) Impact of an artificial surfactant release on air-sea gas fluxes during deep ocean gas exchange experiment II. J Geophys Res 116(C11016). doi:10.1029/2011JC007023

26. Smyth TJ, Tilstone GH, Groom SB (2005) Integration of radiative transfer into satellite models of ocean primary productivity. J Geophys Res 110(C10014). doi:10.1029/2004JC002784

27. Takahashi T, Sutherland SC, Wanninkhof R, Sweeney C, Feely RA, Chipman DW, Burke Hales B, Friederich G, Chavez F, Watson AJ, Bakker DCE, Schuster U, Metzl N, Yoshikawa-Inoue H, Ishii M, Midorikawa T, Sabine C, Hoppema JMJ, Olafsson J, Arnarson TS, Tilbrook B, Johannessen T, Olsen A, Bellerby R, Baar HJWd, Nojiri Y, Wong CS, Delille B (2009)

Climatological mean and decadal change in surface ocean pCO_2 and net sea-air CO_2 flux over the global oceans. Deep Sea Res II 56:554–577

28. Tsai W-T, Liu K–K (2003) An assessment of the effect of sea surface surfactant on global atmosphere-ocean CO_2 flux. J Geophys Res 108(C4). doi:10.1029/2000JC000740

29. Wanninkhof R (1992) Relationship between wind speed and gas exchange over the ocean. J Geophys Res 97(C5):7373–7382

30. Wanninkhof R, McGillis WR (1999) A cubic relationship between air-sea CO_2 exchange and wind speed. Geophys Res Lett 26(13):1889–1892

31. Ward B, Wanninkhof R, McGillis WR, Jessup AT, DeGranpre MD, Hare JE, Edson JB (2004) Biases in the air-sea flux of CO_2 resulting from surface temperature gradients. J Geophys Res 109(C08S08). doi:10.1029/2003JC001800

32. Watson AJ, Schuster U, Bakker DCE, Bates NR, Corbiere A, Gonzalez-Davila M, Friedrich T, Hauck J, Heinze C, Johannessen T, Kortzinger A, Metzl N, Olafsson J, Olsen A, Oschlies A, Padin XA, Pfeil B, Santana-Casiano JM, Steinhoff T, Telszewski M, Rios AF, Wallace DWR, Wanninkhof R (2009) Tracking the variable North Atlantic sink for atmospheric CO_2. Science 326:1391–1393

33. Woolf DK (2005) Parameterization of gas transfer velocities and sea-state dependent wave breaking. Tellus 57B:87–94

34. Yang M, Blomquist BW, Fairall CW, Archer SD, Huebert BJ (2011) Air-sea exchange of dimethylsulfide in the Southern Ocean: measurements from SO Gasex compared with temperate and tropical regions. J Geophys Res 116(C00F05). doi:10.1029/2010JC006526

35. Yang M, Blomquist BW, Huebert BJ (2009) Constraining the concentration of the hydroxyl radical in a stratocumulous-topped marine boundary layer from sea-to-air eddy covariance flux measurements of dimethylsulfide. Atmos Chem Phys 9:9225–9236

ISMER—Active Magmatic Processes in the East African Rift: A Satellite Radar Perspective

Juliet Biggs, Elspeth Robertson and Mia Mace

Abstract An understanding of the fundamental processes by which continental rifts form and develop is necessary to fully understand plate tectonics and the associated earthquake and volcano hazards. Fault slip, magma chamber inflation and magma intrusion produce characteristic patterns of surface deformation which can be the key to characterising the active tectonic and magmatic processes. The ISMER project uses archived and scheduled satellite radar images to analyse surface displacements along the length of the East African Rift over the past 15 years. Where available we use seismological observations from global catalogues and local networks to target our observations, and perform systematic surveys over key areas to identify aseismic processes. In this article, we summarise published findings on the Kenyan (1997–2008) and Main Ethiopian Rifts and the 2009 Malawi earthquakes and describe new results from the Western Branch, the Kenyan Rift (2008–2010) and a series of seismic swarms. Due to InSAR studies such as these, the number of volcanoes known to be active in East Africa has gone from 3 (Nyiragongo, Nyamuragira and Oldonyo Lengai) to > 14 with implications for volcanic hazard and geothermal potential.

Keywords InSAR · Magmatic activity · Volcanoes deformation · Hazard monitoring

J. Biggs (✉)
COMET+, Department of Earth Sciences, University of Bristol, Bristol, UK
e-mail: Juliet.biggs@bristol.ac.uk

E. Robertson
Department of Earth Sciences, University of Bristol, Bristol, UK

M. Mace
Department of Physics, University of Bristol, Bristol, UK

D. Fernández-Prieto and R. Sabia, *Remote Sensing Advances for Earth System Science*, SpringerBriefs in Earth System Sciences, DOI: 10.1007/978-3-642-32521-2_9, © The Author(s) 2013

1 Introduction

Divergent plate boundaries start as continental rifts and mature through a series of distinct stages to form mid-ocean ridges. Established models of continental rift formation assume all extension takes place on faults yet recent observations from the East African Rift have shown the importance of magmatic intrusion [12, 22] and active volcanic systems [6]. However, detailed studies only exist for small areas and a handful of events, the coverage is insufficient to determine the temporal and spatial distribution of tectonic and magmatic processes and their implications for rift development. Now, a systematic study is required to quantify the role magma plays in accommodating extension and its relationship to fault-based stretching. By comparison between mature and immature rift segments, such a survey can not only observe present-day processes but also evaluate their contribution to the development of continental rifts.

The rapidly developing field of satellite geodesy has revolutionised surface deformation measurements, providing new insights into active tectonics, and in particular identifying activity that is not accompanied by significant seismicity. Interferometric Synthetic Aperture Radar (InSAR) compares the phase of radar images taken at different times to detect small (<1 cm) surface displacements at high resolution (<90 m) over continental-scale areas. Satellite-based InSAR requires no ground deployments, making it an ideal tool for detecting and comparing processes that occur over large areas of remote terrain.

The ISMER project uses past and future satellite radar images to analyse surface displacements along the length of the East African Rift over the past 15 years (Fig. 1). This region includes more than 30 volcanoes, few of which have ever been studied for geodetic activity, and several seismic swarms. A large archive of data exists for the Main Ethiopian Rift and Western Branch of the East African Rift. ERS 2 made acquisitions in 1997 and 2000, and there are 3–4 acquisitions per year for Envisat in the time period 2003–2008. From 2008, acquisitions were scheduled in response to events of interest (such as the 2009 Karonga Earthquake Sequence in Malawi) and to improve the temporal resolution of observation over areas of particular interest (e.g. Kenyan volcanoes).

Results from the ISMER project have been published in a series of peer-reviewed articles starting with the preliminary study of the Kenyan Rift from 1992 to 2008 [6], the 2009 Karonga earthquakes, Malawi [8] and a systematic study of the Main Ethiopian Rift [7]. These are summarised briefly alongside new results from the Western Branch and from a program of scheduled Envisat acquisitions on the Kenyan Rift in 2009–2010.

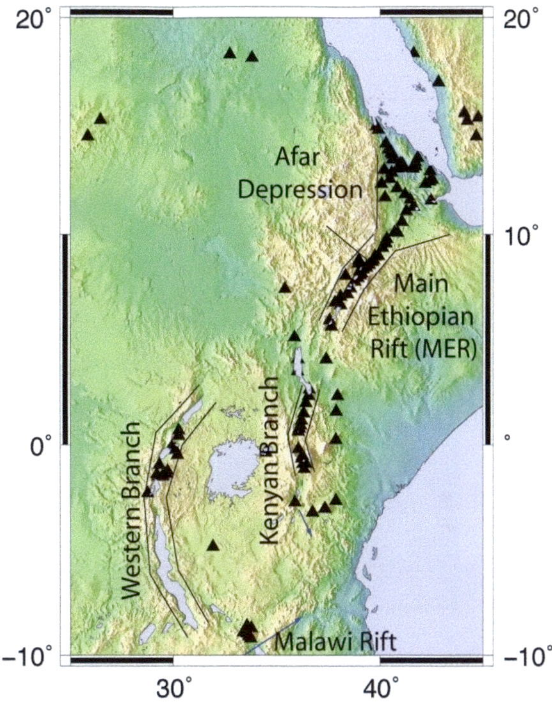

Fig. 1 The East African Rift—a divergent plate boundary between the Nubian and Somalian Plates. In the Afar depression to the north, spreading is accommodated by magmatic processes while the Malawi rift in the south is a tectonically-controlled basin

2 Seismic Activity

Earthquakes occur in response to stress changes in the Earth's upper crust and thus can be used to target observations. Tectonic earthquakes associated with slip on major structural faults would be expected to have a single large magnitude event followed by smaller aftershocks. In contrast, seismic activity associated with magma movement (for example a dyke intrusion) usually consist of a swarm of small earthquakes lasting for days to weeks. Large earthquakes ($M > 4.5$) can be detected using the Global Seismographic Network (GSN), but swarms of smaller earthquakes may only be detected on temporary deployments of local networks. In this section we discuss the Karonga earthquake sequence, which was detected using the GSN, and a series of swarms reported by local networks.

2.1 The 2009 Karonga Earthquake Sequence

The Karonga region on the western shore of Lake Malawi (Fig. 2a) suffered a series of earthquakes in December 2009. The largest four events had magnitudes

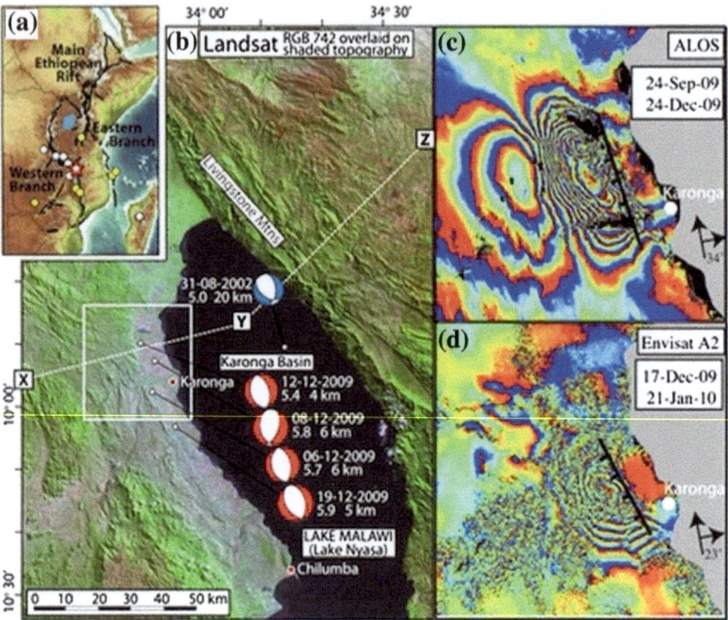

Fig. 2 Interferograms and models for the December 2009 Karonga Earthquake Sequence, Malawi **a**. **b** Location of the 4 largest earthquakes including their focal mechanisms (after [8]). **c** Interferogram spanning the entire earthquake sequence and **d** Interferogram spanning the December 19th Earthquake only

greater than 5.5. According to United Nations reports, over 1000 houses collapsed, a further 2900 were damaged, 300 people were wounded, and 4 were killed. As part of the ISMER project, the European Space Agency (ESA) rapidly scheduled the Envisat to acquire images on all relevant beam modes and tracks in the next few weeks. In this region, archived images were available in several beam modes including D2 and D5 (from May 2009), A4 in September and A3 in December. Due to a swift response, an image was collected in beam mode A2 on December 17th allowing us to isolate the deformation from the 19th December earthquake (Fig. 2d). The coastal plain lost interferometric coherence within a few months but bedrock west of the Karonga fault maintained coherence.

Interferograms spanning all the earthquakes show a fringe pattern elongated NNW-SSE (Fig. 2c). A central lobe shows up to 28 cm of line-of-sight range increase with smaller amounts of range decrease in lobes to the east and west. The boundary between central and eastern lobes is sharp suggesting a fault close to the surface, but the boundary between the western and central lobes is smoother. The interferogram covering the 19th December earthquake (Fig. 2d) shows a simpler bulls eye pattern of 4 fringes, and displays the same sharp discontinuity as obvserved in the longer interferograms.

The most coherent interferograms were used to produce a fault model based on existing analytic solutions and inversion methods [17, 23]. The geology and geomorphology suggest an east-dipping fault plane aligned with a prominent scarp, but this cannot match the sharp discontinuity observed in the surface displacement fields. However, a single W-dipping fault plane extending from the surface discontinuity to a depth of 6 km was found to fit all interferograms well.

The Karonga sequence offers an insight into active rift processes in the East African Rift System. Despite the swarm-like seismicity, we found no evidence for the involvement of magmatic fluids. The deformation is consistent with the rupture of a single west-dipping fault, which appears on geological maps but has little geomorphological expression.

2.2 Seismic Swarms

Seismic swarms occur in a variety of different environments and might have a diversity of origins. They occur frequently in areas of magmatic and volcanic activity, but also associated with aseismic fault slip, particularly at releasing bends on transform faults [13]. Since the intrusion of a shallow dike is typically associated with surface displacements on the order of 0.1–1 m, InSAR is the ideal tool to distinguish the cause. For example, surface deformation associated with the 2007 Lake Natron/Gelei Swarm linked it to the intrusion of a 2.4 m wide dike [1, 3, 5, 9]. Using archived data from the background mission, can be used to reassess past activity; the 2000 Ayelu-Amoissa swarm was recently attributed to an intrusion of a 1.5 m wide dike [15].

Compared to other parts of the East African Rift System, the Kenyan branch of the EAR has low seismic activity with no events $M > 5$ recorded since 1928. However, local networks frequently report high levels of microseismicity with $10M < 3$ events per day in the southernmost part of the rift [14]. This microseismicity is often clustered in both space and time forming microseismic swarms.

We form interferograms spanning the 1998 Lake Magadi swarm [14] and the Lake Manyara swarm [1, 16]. The Lake Manyara swarm appears to be continuously active for at least 12 years as it was detected by local deployments in 1994–1995 [16] and in 2007 [1] with consistent locations, magnitudes and event rates. The Lake Magadi region appears to have some background levels of micro-seismic activity, but a jump from 10 to 300 events per day was detected in April 1998. The b-value of 0.75 suggests a tectonic origin rather than magmatic origin [14]. Both swarms are elongated NE-SW sub-parallel to the trend of the rift axis.

Figure 3 shows interferograms spanning the 1998 Lake Magadi Swarm, with seismicity from [14] and the Lake Manyara Seismic swarm with seismicity from the 2007 temporary network [1]. The Lake Magadi interferogram show a strong atmospheric influence, but neither show sufficient deformation to be attributed to a dike intrusion. We conclude that these swarms were not driven by magmatic

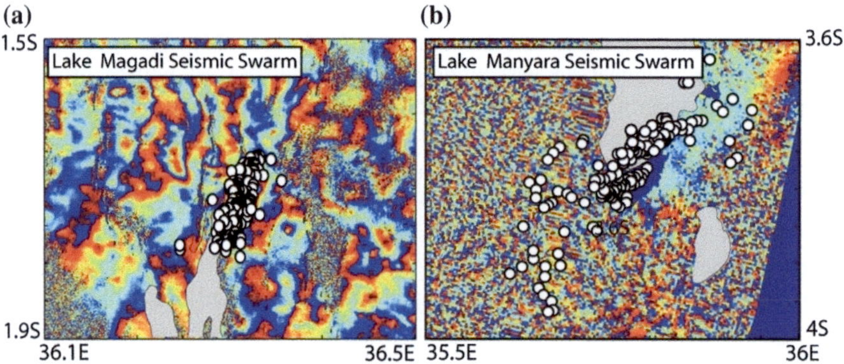

Fig. 3 Interferograms spanning seismic swarms. **a** ERS1/2 interferogram from 23rd September 1997 to 13th September 2000 spanning the Lake Magadi Swarm, with seismicity from [12]; **b** Envisat interferogram of Lake Manyara from 5th March 2007 to 18th February 2008 with seismicity from the 2007 temporary network [1]. The Lake Manyara interferogram was processed at 16 looks and filtered three times in order to improve coherence

activity and must be attributed to either fault slip or the migration of aqueous fluids.

3 Aseismic Deformation

While seismic activity may be a good indicator of activity, many processes that produce surface deformation are essentially aseismic, or associated with seismicity too small to be detected by the available networks. Examples include fault creep, visco-elastic relaxation, magma chamber inflation, geothermal fluid flow and water pumping. Here we follow the approach of Pritchard and Simons, 2004 [18] and perform a systematic survey to detect aseismic deformation.

3.1 Kenyan Rift

A previous InSAR survey of the Kenyan rift discovered four volcanoes underwent geodetic activity between 1997 and 2007 [6]. During the ISMER project, Envisat acquisitions were scheduled from 2008 to monitor any further volcanic deformation. Here we present a summary of deformation from 2008 to 2010 from C-band Envisat and L-band ALOS (an ESA third party mission). Longer wavelengths interact with scatterers that are more stable in time. This causes vegetation to have a greater decorrelating effect at C-band than at L-band.

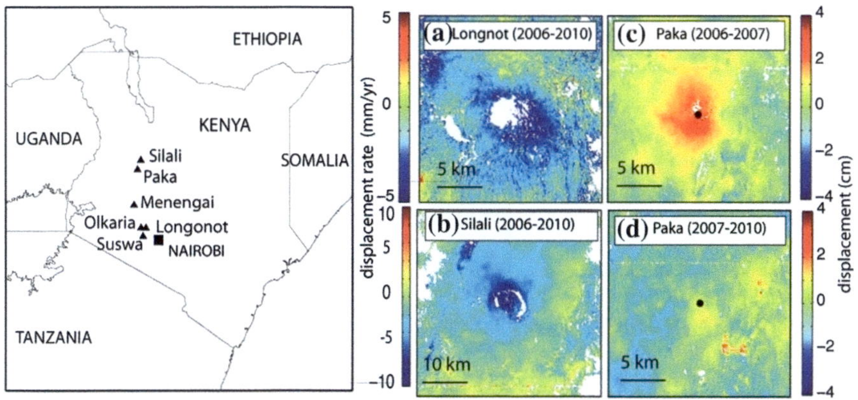

Fig. 4 Summary of observations of the Kenyan Rift during 2007–2010 using 3rd Party JAXA satellite ALOS. An earlier InSAR survey had detected pulses of deformation at Longonot, Suswa, Paka and Menengai volcanoes from 1997 to 2000. **a** Subsidence at Longonot volcano at a rate of ∼2 mm/yr, **b** subsidence at Silali volcano at a rate of 9 mm/yr, **c** uplift of Paka volcano consistent with the end of the previously detected inflation episode lasting until July 2008 **d** period following July 2008 showing no deformation

Longonot volcano inflated by 9.2 cm between 2004 and 2006 [6] but subsequent Envisat images up to 2008 were incoherent. During 2008–2010, Longonot subsided by 0.72 cm at a constant rate of ~ -2 mm per year (Fig. 4a). Paka inflated by 21.3 cm over a 9 month period in 2006–2007 [6]. The ALOS dataset captures the end of this inflation episode. The acquisitions are divided into two time periods: a stack of 4 interferograms from January 2007 to July 2008 show ∼3 cm of uplift (Fig. 4c), while a stack of 6 interferograms after this date show no deformation (Fig. 4d). Silali showed no deformation in the previous InSAR survey but a stack of 11 ALOS interferograms from 2007 to 2010 show shallow subsidence centered on the volcano caldera of up to 4 cm (Fig. 4b).

Suswa and Menengai volcanoes subsided by 4.6 and 3 cm respectively between 1997 and 2000; yet no deformation has been observed since. The Olkaria volcanic complex hosts Kenya's largest geothermal resources with four power plants currently in operation within the volcanic field. Both Alos and Envisat have continuous coherence from 2006 and no ground deformation is observed.

3.2 Main Ethiopian Rift

The Main Ethiopian Rift (MER) contains 13 named volcanoes [19]. Despite numerous strands of geophysical evidence which suggest that there is a plentiful supply of magma in the mantle and crust [4], very few eruptions have been recorded historically. We performed a systematic study using ERS and Envisat

Fig. 5 Volcanoes of Main Ethiopian Rift (after [7]). a Uplift at Alutu (17 December 2003 to 18 August 2004), and b subsidence at Corbetti (23 September 1997 to 13 September 2000). Volcanoes from the Smithsonian Database are denoted with *black triangles H*, Hertali; *D*, Dofen; *F*, Fentale; *B*, Beru; *K*, Kone; *B*, Boset; *S*, Sodore; *G*, Gedemsa; *Bor*, Bora Berricio; *TM*, Tulla Moje; *EZ*, East Ziway; *A*, Alutu; *C*, Corbetti

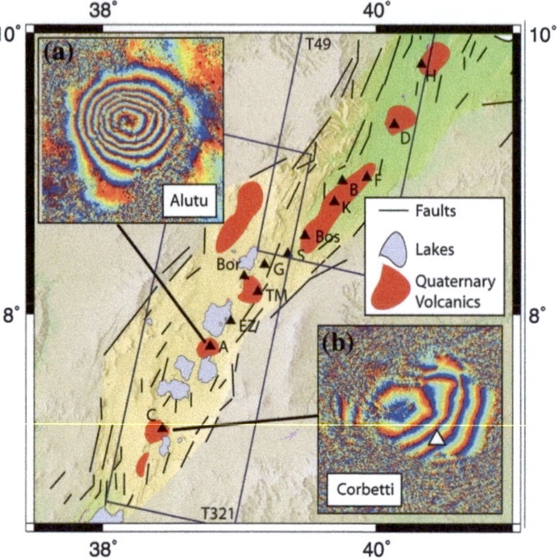

data from 1997 to 2009 which found surface displacements at 4 locations within the rift. Further details are given in Biggs et al. [7].

The volcano Alutu shows significant deformation in all Envisat interferograms processed from December 2003 to March 2009. The deformation pattern is circular and centered on the volcano (Fig. 5a) Time series analysis shows two pulses of rapid inflation (20 cm) in 2004 and 2008 separated by gentle subsidence at a rate of 3–5 cm/yr. Alutu is the site of a geothermal field (the Alutu-Langano Field) which is in the process of being exploited for power production. The combination of subsidence and uplift suggests the complex interaction in a coupled magmatic-geothermal system.

The intereferogram from 1997 to 2000 shows subsidence of at least 11 cm centered on Corbetti volcano (Fig. 5b). Envisat interferograms from 2003 to 2009 show no deformation but uplift restarted in 2010. The deformation corresponds to a period (1996–1998) of intense ground cracks in the nearby areas of Muleti, Lake Shala and Adamitulu. Previous studies [2] found no link to active faulting or distant earthquakes, considered the rates of groundwater withdrawal too low to result in fissures of this kind. Ultimately they conclude that the fissures are created by heavy rainfall, but the InSAR observations suggest that they may be related to volcano deformation.

Deformation of 3–5 cm occurs $\sim$ 10 km south of the volcano Hertali, in an area with an unnamed volcanic edifice; recent-looking lava flows are visible in optical imagery. We have named the volcanic edifice Haledebi following the name for the area on Ethiopian topographic maps. The deformation pattern is sharply bounded on the western side, suggesting that the deformation is fault-controlled, or involves intrusion into a dipping structure.

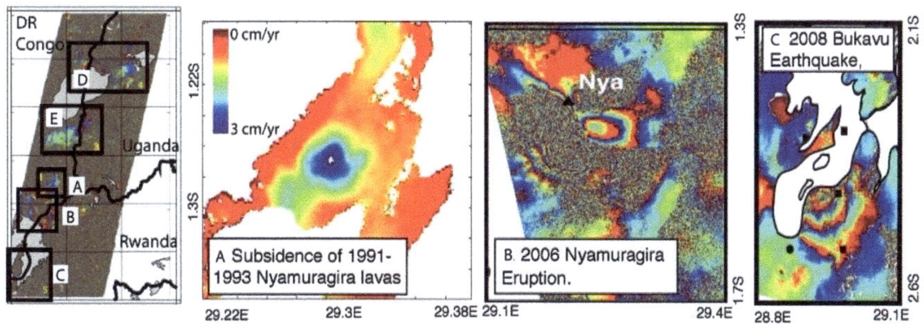

Fig. 6 Envisat Survey of the Western Branch. *A* Ongoing subsidence of the 1991–1993 lava flows of Nyiragongo at 3 cm/yr. *B* Deformation associated with the 2006 Nyamuragira eruption, *C* deformation associated with the 2008 Bukavu Earthquake. *D* Shows an area of coherence close to the volcano May-ya-moto; *E* shows an area of coherence around the Bunyaruguru and Katwe-Kikorongo volcanic fields. Neither show any deformation during the period of the survey

Together, these observations indicate a shallow (<10 km), frequently replenished zone of magma storage associated with volcanic edifices and add to the growing body of observations that indicate shallow magmatic processes operating on a decadal timescale are ubiquitous throughout the East African Rift. In the absence of detailed historical records of volcanic activity, satellite-based observations of monitoring parameters, such as deformation, could play an important role in assessing volcanic hazard.

3.3 Western Branch

The western branch of the East African rift is both more seismically and more volcanically active than the Kenyan branch. Previous InSAR studies have focussed on the eruptions of Nyamuragira and Nyiragongo and earthquakes with $M > 5.5$. Here we perform a systematic InSAR survey of all volcanoes from Lake Kivu to north of Lake Edward using Envisat data from 2005 to 2010. However, due to the effect of dense vegetation on C-band radar, the majority of the region was incoherent. The only coherent regions are the Bunyaruguru, Katwe-Kikorongo and Nyiragongo-Nyamuragira volcanic fields and the area near May-ya-moto volcano.

Short-term deformation was detected associated with the 2006 eruption of Nyamuragira (Fig. 6b) as reported by [10] and the 2008 Bukavu earthquakes (Fig. 6c) as reported by [11]. Subsidence was detected in an area to the north of Nyamuragira at a constant rate of 3 cm/yr (Fig 6a). The area of subsidence corresponds with the lava flows from 1991 to 1993 [20] and can be attributed to the cooling, contraction and repacking of the lava [21].

4 Conclusions

The ISMER project has performed a systematic survey of tectonically- and magmatically-driven surface deformation along the major sections of the East African Rift. It has continued monitoring of the Kenyan Rift, allowing us to detect slower deformation (mm/yr) such as at Longonot and Silali volcanoes; it has performed new surveys of the Main Ethiopian Rift and the Western Branch of the rift revealing 4 deforming areas, including cycles of uplift and subsidence at the Alutu-Langano geothermal field, and a previously unnamed volcanic edifice. The new acquisition strategy has enabled us to respond to seismic events, such as the 2009 Karonga earthquake sequence in Malawi and the archived background mission data has been used to look for surface deformation associated with published seismic swarms.

Acknowledgments This project was supported by the ESA Changing Earth Science Network, COMET+ at the University of Oxford and a lectureship at the University of Bristol. ER is supported by a NERC studentship to the University of Bristol; MM was supported by a NERC undergraduate summer placement. With thanks to Diego Fernandez, Steffen Dransfeld, Roberto Sabia and many other people at ESA who have helped in so many ways. We thank the anonymous reviewer for helpful comments. Much of the work summarised here has been published in other articles and I thank my co-authors: Ian Bastow, Derek Keir, Elias Lewi, Edwin Nissen, Tim Craig, James Jackson and David Robinson.

References

1. Albaric J, Perrot J, Déverchère J, Deschamps A, Le Gall B, Ferdinand RW, Petit C, Tiberi C, Sue C, Songo M (2010) Contrasted seismogenic and rheological behaviours from shallow and deep earthquake sequences in the North Tanzanian Divergence, East Africa. J Afr Earth Sci 58:799–811
2. Ayalew L, Yamagishi H, Reik G (2004) Ground cracks in Ethiopian rift valley: facts and uncertainties. Eng Geol 75:309–324
3. Baer G, Hamiel Y, Shamir G, Nof R (2008) Evolution of a magma-driven earthquake swarm and triggering of the nearby Oldoinyo Lengai eruption, as resolved by InSAR, ground observations and elastic modeling, East African Rift, 2007. Earth Planet Sci Lett 272:339–352
4. Bastow I, Keir D, Daly E (2011) The Ethiopia Afar Geoscientific Lithospheric Experiment (EAGLE): Probing the transition from continental rifting to incipient seafloor spreading. In: Beccaluva L, Bianchini G, Wilson M (eds) Volcanism and evolution of the African lithosphere, vol 478. GSA Special Paper, Boulder, pp 1–26. doi:10.1130/2011.2478(04)
5. Biggs J, Amelung F, Gourmelen N, Dixon TH, Kim S (2009) InSAR observations of 2007 Tanzania rifting episode reveal mixed fault and dyke extension in an immature continental rift. Geophys J Int 179:549–558. doi:10.1111/j.1365-246X.2009.04262.x.
6. Biggs J, Anthony E, Ebinger C (2009) Multiple inflation and deflation events at Kenyan volcanoes, East African Rift. Geology 37:979–982. doi:10.1130/G30133A.1
7. Biggs J, Bastow ID, Keir D, Lewi E (2011) Pulses of deformation reveal frequently recurring shallow magmatic activity beneath the Main Ethiopian Rift. Geochem Geophys Geosyst 12:Q0AB10. doi:10.1029/2011GC003662

8. Biggs J, Nissen E, Craig T, Jackson J, Robinson DP (2010) Breaking up the hanging wall of a rift-border fault: the 2009 Karonga earthquakes, Malawi. Geophys Res Lett 37:L11305. doi:10.1029/2010GL043179

9. Calais E, d'Oreye N, Albaric J, Deschamps A, Delvaux D, Dverchre J, Ebinger C, Ferdinand R, Kervyn F, Macheyeki A, Oyen A, Perrot J, Saria E, Smets B, Stamps D, Wauthier C (2008) Strain accommodation by slow slip and dyking in a youthful continental rift, East Africa. Nature 456:783–787. doi:10.1038/nature07478.

10. D'Oreye N, Fernandez J, Gonzalez P, Kervyn F, Wauthier C, Frischknecht C, Calais E, Heleno S, Cayol V, Oyen A, Marinkovic P (2008) Systematic insar monitoring of african active volcanic zones: what we have learned in three years, or an harvest beyond our expectations. In: Second workshop on use of remote sensing techniques for monitoring volcanoes and seismogenic areas, 2008, USEReST 2008, pp 1–6. DOI:10.1109/USEREST.2008.4740361

11. D'Oreye N, González PJ, Shuler A, Oth A, Bagalwa L, Ekström G, Kavotha D, Kervyn F, Lucas C, Lukaya F, Osodundu E, Wauthier C, Fernández J (2011) Source parameters of the 2008 Bukavu-Cyangugu earthquake estimated from InSAR and teleseismic data. Geophys J Int 184:934–948. doi:10.1111/j.1365-246X.2010.04899.x.

12. Grandin R, Socquet A, Binet R, Klinger Y, Jacques E, de Chabalier J, King GCP, Lasserre C, Tait S, Tapponnier P, Delorme A, Pinzuti P (2009) September 2005 Manda Hararo-Dabbahu rifting event, Afar (Ethiopia): constraints provided by geodetic data. J Geophys Res 114(B13):B08,404. doi:10.1029/2008JB005843

13. Holtkamp SG, Pritchard ME, Lohman RB (2011) Earthquake swarms in South America. Geophys J Int 187:128–146. doi:10.1111/j.1365-246X.2011.05137.x.

14. Ibs-von Seht M, Plenefisch T, Klinge K (2008) Earthquake swarms in continental rifts. A comparison of selected cases in America, Africa and Europe. Tectonophysics 452:66–77. doi:10.1016/j.tecto.2008.02.008

15. Keir D, Pagli C, Bastow ID, Ayele A (2011) The magma-assisted removal of Arabia in Afar: evidence from dike injection in the Ethiopian rift captured using InSAR and seismicity. Tectonics 30:TC2008. doi:10.1029/2010TC002785.

16. Mulibo G, Nyblade A (2009) The 1994–1995 manyara and kwamtoro earthquake swarms: variation in the depth extent of seismicity in northern tanzania. S Afr J Geol 112(3–4):387–404. doi:10.2113/gssajg.112.3-4.387

17. Okada Y (1985) Surface deformation due to shear and tensile faults in a half-space. Bull Seismol Soc Am 75:1135–1154

18. Pritchard ME, Simons M (2004) An InSAR-based survey of volcanic deformation in the central Andes. Geochem Geophys Geosyst 5:2002. doi:10.1029/2003GC000610.

19. Siebert L, Simkin T (2002) Volcanoes of the world: an illustrated catalog of Holocene volcanoes and their eruptions. Smithsonian Institution, Global Volcanism Program Digital Information Series: http://www.volcano.si.edu/world/

20. Smets B, Wauthier C, d'Oreye N (2010) A new map of the lava flow field of Nyamulagira (D.R. Congo) from satellite imagery. J Afr Earth Sci 58:778–786

21. Toombs A, Wadge G (2012) Co-eruptive and inter-eruptive surface deformation measured by satellite radar interferometry at Nyamuragira volcano, D.R. Congo, 1996 to 2010. J Volcanol Geoth Res 245:98–122. doi:10.1016/j.jvolgeores.2012.07.005.

22. Wright T, Ebinger C, Biggs J, Ayele A, Yirgu G, Keir D, Stork A (2006) Magma-maintained rift segmentation at continental rupture in the 2005 Afar dyking episode. Nature 442:291–294

23. Wright T, Parsons B, Jackson J, Haynes M, Fielding E, England P, Clarke P (1999) Source parameters of the 1 October 1995 Dinar (Turkey) earthquake from SAR interferometry and seismic bodywave modelling. Earth Planet Sci Lett 172:23–37

FEMM—Fire Effects Modelling and Mapping: An Approach to Estimate the Spatial Variability of Burning Efficiency

Patricia Oliva

Abstract Burning efficiency, defined as the percentage of biomass consumed by the fire, plays a key role in the estimation of the amount of gases released from the biomass burning process. Traditionally this parameter was assigned to a vegetation class as a constant value. However, different levels of damage, also called burn severity, may occur within an area affected by fire. This affects the amount of biomass consumed by burning. Consequently, the burning efficiency coefficients should vary accordingly to the levels of damage presented in the burned area. The approach described in this study estimates the burning efficiency values adjusting the burning efficiency coefficients by vegetation type found in the literature to the level of damage computed from MERIS images. This study focused in two big fires occurred in Spain in 2009. The burning efficiency maps generated highlighted the overestimation produced when the level of damage is not considered in the burning efficiency estimation.

Keywords Burning efficiency · Level of damage · Forest fires · Emissions · Biomass burning · Forest fires burn severity

1 Introduction

Forest fires seriously affect the ecosystem dynamics by consuming the vegetation layer and modifying the soil structure [1]. Besides, the biomass burning process releases greenhouse gases (GHG), trace gases and aerosols to the atmosphere [2]

P. Oliva (✉)
Department of Geography, University of Alcala,
Colegios 2 28801 Alcala de Henares, Spain
e-mail: patricia.oliva@uah.es

D. Fernández-Prieto and R. Sabia, *Remote Sensing Advances for Earth System Science*,
SpringerBriefs in Earth System Sciences, DOI: 10.1007/978-3-642-32521-2_10,
© The Author(s) 2013

which play a decisive role in the climate change. Consequently, special attention has been given by the scientific community to accurately estimate the amount of gases released from biomass burning [3].

In this study we followed the equation proposed by Seiler and Crutzen [4],

$$M_{i,j,k} = BL_{i,j,m} \times BE_{i,j,m} \times BS_{i,j} \times EF_k \times 10^{-15}$$

where $M_{i,j,k}$ is the amount of gas k released for a specific area (coordinates i, j) in teragrams, $BL_{i,j,m}$ is the biomass load (dry matter) of an homogeneous fuel or vegetation type m for the same area (g/m^2), $BE_{i,j,m}$ is the burning efficiency of the fuel or vegetation type m (0–1, dimensionless), $BS_{i,j}$ is the burned surface (m^2), and EF is the emission factor of the gas k (g/kg of dry matter).

These parameters have several sources of uncertainties which hinder their accurate estimation [5]. Significant improvement has done in order to reduce these uncertainties. However, it is necessary to study in detail the spatial distribution of burning efficiency in order to improve the current estimations.

Burning efficiency (BE) is defined as the percentage of biomass consumed by the fire. Traditionally, the emission estimates assign a BE value for each vegetation or fuel type, considering that the vegetation affected by the fire was completely burnt [5]. However, some studies concluded that BE might be considered as a spatially dynamic parameter rather than constant, as the current BE factors led to uncertainties ranging from 23 to 46 % [6, 7].

In this study we propose an approach to estimate the burning efficiency from satellite images. In order to capture the variability of this parameter we consider the level of damage produce by the fire by measuring the burn severity (BS). BS, defined as the degree of damage on soil and plants after the fire is extinguished [8, 9], offers an estimation of the biomass consumed by the fire. Therefore, we used a BS map to adapt the BE coefficients found in the literature to the spatial-variability of the actual BE.

2 Methodology

2.1 Study Area and Satellite Data

Two large fires occurred during the 2009 fire season in Spain were selected for this study because they showed a wide range of severity levels within them. The first one was declared on the 28th of July 2009 near Arenas de San Pedro (Avila) (Fig. 1). A total surface of 3920 ha was burned, mainly composed by pine forest and dense shrublands.

The second forest fire occurred in Aliaga (Teruel) burning almost 6315 ha of pine forests, mixed forests, and shrublands. This fire started on the 22nd of July by lightning and was completely extinguished after 4 days.

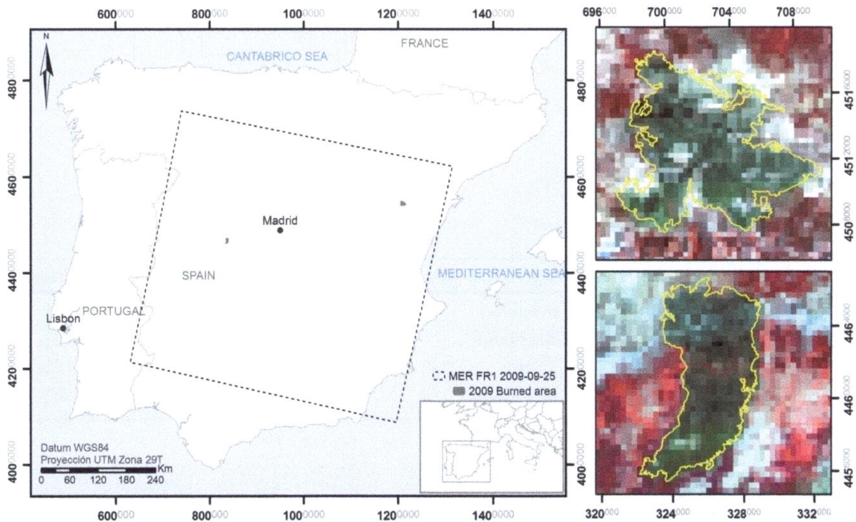

Fig. 1 Location of the forest fires selected in the 2009 fire season. The forest fires are displayed on the right side. Fire number 1 refers to Teruel's fire and number two is Avila's fire. A MERIS image dated on the 25th of September is used as background (colour composite 13/8/5)

The data used in this study were acquired by the Medium Resolution Imaging Spectrometer (MERIS) sensor, one of the ten instruments on board the Envisat (Environmental Satellite) of the European Space Agency. MERIS measures the solar radiation reflected by the Earth at a ground spatial resolution of 300 m in 15 spectral bands between 390 and 900 nm [10].

A Full Resolution-Level 1b image dated on 25th September 2010 was used in this study. This product offers top of atmosphere reflectance of the 15 spectral bands. The image was atmospherically corrected applying the SCAPE-M routine [11].

2.2 Burning Efficiency Estimation

2.2.1 Workflow

The burning efficiency estimation developed in this approach considered three types of inputs: BE coefficients from previous studies, a land cover or vegetation map, and a map of level of damage produced by the fire (Fig. 2). Therefore, the BE coefficients were adjusted to the vegetation classes and level of damage.

Fig. 2 Workflow of the burning efficiency estimation developed in this project

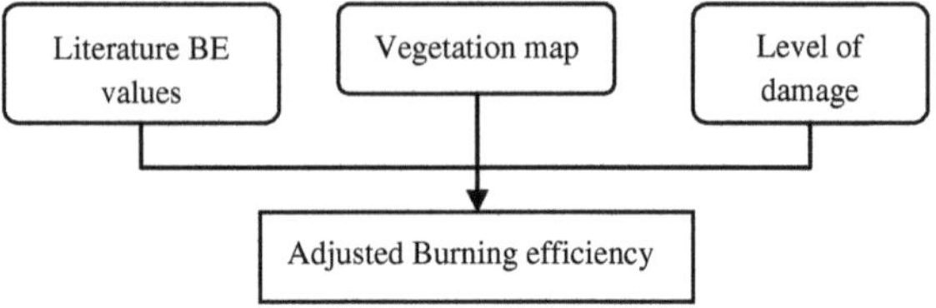

Table 1 Burning efficiency values from previous studies by type of vegetation

Vegetation type	BE_{Lit}	Reference
Grassland	0.98	[12]
Shrubland	0.95	[13]
Conifer	0.57	[14]
Deciduous	0.56	[14]

2.2.2 Burning Efficiency Coefficients

After a thorough search of BE coefficients, we selected the coefficients which matched the conditions of the Mediterranean forested areas (Table 1). Hereafter these coefficients are referred to as BE_{Lit}.

2.2.3 Vegetation Map

The vegetation categories used in previous studies are very diverse since they adapted the categories to the ecosystem they were studying or to the land cover map they could obtain. In our case, to define the vegetation categories which fit the BE_{Lit} coefficients we selected the Globcover map. This map complies with the requirements of this study because of two reasons. First, the Globcover map was derived from MERIS images. Second, the vegetation classes of this map were described with more detail than other land cover maps. In order to match the vegetation map with the BE coefficients, the vegetation categories were simplified to grassland, shrubland, conifers (evergreen) and deciduous.

2.2.4 Burn Severity Estimation

Burn severity (BS) was estimated by means of the simulation model developed by De Santis et al. [15]. These authors proposed the inversion of two linked Radiative Transfer Models (RTMs), PROSPECT and GeoSail. The model was composed by a *Look-up-table* of 30 spectra corresponding to GeoCBI (Geophysical Composite Burn Index) values ranging from 0 to 3 [16]. These spectra were organized as a spectral library which provided reference spectra to run a spectral angle mapper supervised classification [17] of a single post-fire image. The result was a burn severity map, in which a corresponding GeoCBI value was assigned to

Table 2 Burning efficiency values adjusted to level of damage

Vegetation type	BE adjusted to level of damage		
	Low	Medium	High
Grassland	0.85	0.9	0.98
Shrubland	0.7	0.85	0.95
Needleleaved forest	0.25	0.42	0.57
Broadleaved forest	0.25	0.4	0.56

each pixel. This model was applied in Landsat-TM and MERIS data. Landsat-TM data were validated with field measures and were used then to validate the MERIS data [18].

The validation results of the MERIS burn severity estimation showed values of the coefficient of determination higher than 0.92 with a slope of the regression line higher than 0.9 [18]. Those results proved the potential ability of MERIS data to estimate burn severity levels.

2.2.5 Adjusted Burning Efficiency

In this first approach to burning efficiency estimation we followed the methodology proposed by Kasischke et al. [19]. Therefore, the burn severity levels were grouped into three categories which defined the level of damage produced by the fire as low, medium and high. The BE_{Lit} values were adapted to the levels of damage taking into account the following assumptions:

- Low damage: BS values lower than 2.5 means low damage to the tree cover, the shrubland is scorched and the ground is consumed.
- Medium damage: BS values between 2.5 and 2.8 are produced when ground and Shrubland strata are completely consumed by the fire, and the lower braches of the trees are burned but the higher branches remain unaffected.
- High damage: BS values higher than 2.8 means leaves and branches of the tree cover are severely affected by the fire. Shrubland and ground are completely consumed.

Minimum and maximum BE_{Lit} values were assigned to the low and high damage classes, respectively. The BE value for the medium damage class was computed by interpolation from the previous values (Table 2).

3 Results

The BE_{Lit}, the vegetation map and the map of level of damage were combined to obtain the BE adjusted estimation (hereafter BE_{Adj}) (Figs. 3 and 4).

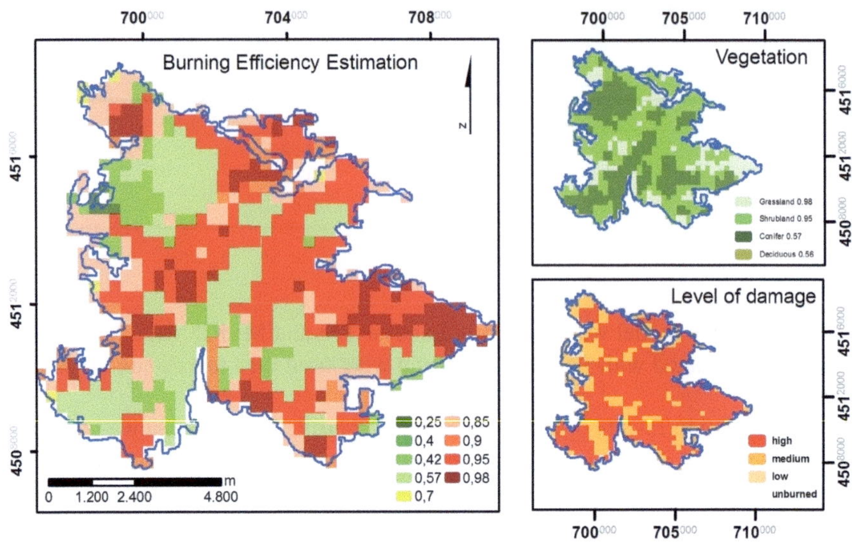

Fig. 3 Teruel's fire burning efficiency estimation adjusted to the vegetation type (*top right*) and the level of damage (*bottom right*)

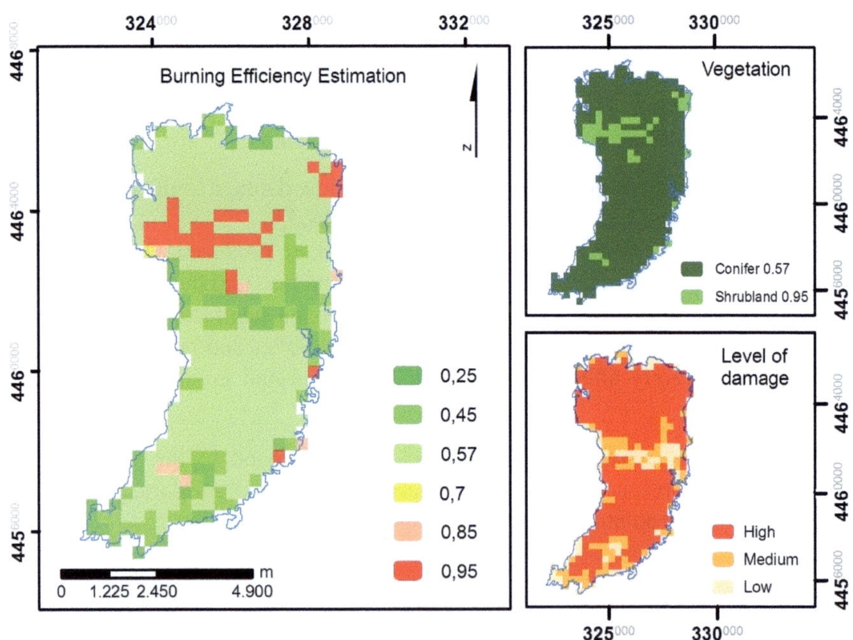

Fig. 4 Avila's fire burning efficiency estimation adapted to the vegetation type (*top right*) and the level of damage (*bottom right*)

Table 3 Comparison between BE coefficients found in literature and BE coefficients adjusted to the level of damage. The rate of change refers to the percentage of change that the mean of the adjusted BE represents in relation with the BE Lit

	Teruel			Avila		
	BE_{Lit}	Mean BE_{Adj}	Rate of change (%)	BE_{Lit}	Mean BE_{Adj}	Rate of change (%)
Conifer	0.57	0.51	-10	0.57	0.47	-17.5
Shrubland	0.95	0.90	-5.3	0.95	0.91	-4.2
Grassland	0.98	0.95	-3			
Deciduous	0.56	0.40	-28.6			

In Teruel the level of damage was high in most of the burned area so the variability of BE_{Adj} coefficients was induced mainly by the vegetation type affected. On the other hand the variability of BE_{Adj} coefficients found in Avila's fire was caused mainly by the level of damage as the vegetation map is dominated by the conifer category.

4 Discussion

Biomass consumption within a burned area is affected by weather conditions, fuel moisture content, vegetation type, and terrain characteristics [5]. Consequently, the burning efficiency coefficient, which measures the amount of biomass consumed by the fire, is influenced by the same factors. However, the spatial variability of the BE coefficients within a vegetation type is not taken into consideration in most of the GHG emission estimations.

The simple approach applied in this study obtained a BE_{Adj} map which showed the increased variability of the BE values when the level of damage is considered in the BE estimation.

In order to quantify the actual differences derived from the use BE_{Adj} values instead of the BE_{Lit} values, we computed the mean and the rate of change of the BE coefficients by vegetation type in both study sites (Table 3). The rate of change is computed as the difference between the BE_{Adj} and the BE_{Lit} divided by the BE_{Lit}. All the rates of change obtained were negative which indicated that the BE_{Adj} offered lower values than the BE_{Lit} coefficients. In other words, the BE_{Lit} coefficients may be overestimating the burning efficiency by 10–28 %. This estimation supported the results obtained in previous studies which attempted to estimate the uncertainties related to the BE coefficient in the estimation of GHG emissions released from the biomass burning process [6, 7, 20].

On the other hand, the higher rate of change is centred in the vegetation types dominated by trees. Conifer and deciduous categories showed lower values of level of damage which indicated that these vegetation types are less affected by fire than the shrubland and grassland categories. This effect is caused by the size and amount of the biomass type in each vegetation category. The leaves, thin branches

(<30 cm) and litter are more sensitive to fire and normally present high values of BE. Whereas trunks and bigger branches show low BE values [21, 22]. As a high percentage of biomass in conifer and deciduous forest is formed by trunks and big branches the BE values of these land covers are lower then shrubland and grassland. Besides, as Mediterranean fires are mainly ground fires the higher crowns are not affected by the fire which reduces the BE value.

5 Conclusions

Traditionally, the emissions estimates assume an average value of BE for each vegetation or fuel type, considering the vegetation affected as completely burnt. However, the different burn severity levels within a fire make evident that the BE should be a dynamic factor instead. The approach presented in this paper used the burn severity values as an estimation of the level of damage produced by the fire to adjust the burning efficiency coefficients found in the literature.

The BE map generated highlighted the spatial variability of this parameter within a vegetation type. The rate of change between the BE_{Lit} and the BE_{Adj} was higher in the conifer and deciduous categories. Therefore, the BE_{Lit} of shrubland and grassland might not need to be adjusted as the level of damage they suffer is normally high. However, the BE_{Lit} values of conifers and deciduous forest should not be used without taking into consideration the level of damage because it may lead to significant overestimations.

References

1. Doer SH, Shakesby RA, Blake WH, Chafer CJ, Humfreys GS, Wallbrink PJ (2006) Effect of differing wildfire severities on soil wettability and implications for hydrological response. J Hydrol 319:295–311
2. Andreae MO, Merlet P (2001) Emission of trace gases and aerosols from biomass burning. Global Biogeochem Cycles 15(4):955–966
3. Vermote E, Ellicot E, Dubovik O, Lapyonok T, Chin M, Giglio L, Roberts GJ (2009) An approach to estimate global biomass burning emissions of organic and black carbon from MODIS fire radiative power. J Geophys Res 114(D18205):23
4. Seiler W, Crutzen PJ (1980) Estimates of gross and net fluxes of carbon between the biosphere and the atmosphere from biomass burning. Clim Chang 2:207–247
5. Palacios-Orueta A, Parra A, Chuvieco E, Carmona-Moreno C (2004) Remote sensing and geographic information systems methods for global spatio-temporal modelling of biomass burning emissions: assessment in the African continent. J Geophys Res 109(D14S09):12
6. Ward DE, Hao WM, Susott RA, Babbitt RE, Shea RW, Kauffmann JB, Justice CO (1996) Effect of fuel composition on the combustion efficiency and emission factors for African savanna ecosystems. J Geophys Res 101(D19):23569–23576
7. Sá ACL, Pereira JMC, Silva JMN (2005) Estimation of combustion completeness based on fire-induced spectral changes in a dambo grassland (Western province Zambia). Int J Remote Sens 26(19):4185–4195

8. Jain T, Pilliod D, Graham R (2004) Confused meanings for common fire terminology can lead to fuels mismanagement. A new framework is needed to clarify and communicate the concepts. Wildfire 4:22–26

9. Key CH, Benson N (2005) Landscape assessment: ground measure of severity, the composite burn index; and remote sensing of severity, the normalized burn ratio. In: Lutes DC, Keane RE, Caratti JF, Key CH, Benson NC, Gangi LJ (eds) FIREMON: fire effects monitoring and inventory system, vol 2004. USDA Forest Service, Rocky Mountain Research Station, General Technical Report. RMRS-GTR-164, Ogden, UT, pp CD:LA1–LA51

10. Bézy JL, Delwart S, Rast M (2000) MERIS—a new generation of ocean-color sensor on board Envisat. ESA Bulletin 103:48–56

11. Guanter L, Gómez-Chova L, Moreno J (2008) Coupled retrieval of aerosol optical thickness, columnar water vapor and surface reflectance maps from Envisat/MERIS data over land. Remote Sens Environ 112:2898–2913

12. Wiedinmyer C, Quayle B, Geron C, Belote A, McKenzie D, Zhang X (2006) Estimating emissions from fires in North America for air quality modelling. Atmos Environ 40:3419–3432

13. IPCC (2006) Volume 4: agriculture, forestry and other land use. In: Eggleston S, Buendia L, Miwa M, Ngara T, Tanabe K (eds) 2006 IPCC guidelines for national greenhouse gas inventories, vol 4. Institute for Global Environmental Strategies (IGES), Hayama, Japan

14. Deeming J, Burgan R, Cohen J (1977) The national fire-danger rating system. U.S. Department of Agriculture, Forest Service, Ogden

15. De Santis A, Chuvieco E, Vaughan PJ (2009) Short-term assessment of burn severity using the inversion of PROSPECT and GeoSail models. Remote Sens Environ 113:126–136

16. De Santis A, Chuvieco E (2009) GeoCBI: a modified version of the composite burn index for the initial assessment of the short-term burn severity sensed data. Remote Sens Environ 113:554–562

17. Debba P, van Ruitenbeek FJA, van der Meer FD, Carranza JM, Stein A (2005) Optimal field sampling for targeting minerals using hyperspectral data. Remote Sens Environ 99:373–386

18. Oliva P, De Santis A (2010) Burn severity estimation using MERIS full resolution imagery. In: Lacoste-Francis H (ed) ESA Living Planet symposium, Bergen, Norway, 28 June—2 July 2010. ESA Communications, Noordwijk, p 8

19. Kasischke ES, French NHF, Bourgeau-Chavez LL, Christensen Jr NL (1995) Estimating release of carbon from 1990 and 1991 forest fires in Alaska. J Geophys Res 100(D2):2941–2951

20. Reid JS, Koppmann R, Eleuterio DP (2005) A review of biomass burning emissions part II: intensive physical properties of biomass burning particles. Atmos Chem Phys 5:799–825

21. Fearnside PM, Graça PMLA, Rodrigues FAJ (2001) Burning of Amazonian rainforest: burning efficiency and charcoal formation in forest cleared for cattle pasture near Manaus, Brazil. For Ecol Manage 146:115–128

22. Araújo TM, Carvalho JA, Higuchi N, Brasil ACP, Mesquita ALA (1999) A tropical rainforest clearing experiment by biomass burning in the state of Pará, Brazil. Atmos Environ 33:1991–1998

Index

D. Fernández-Prieto and R. Sabia, *Remote Sensing Advances for Earth System Science*, 103
SpringerBriefs in Earth System Sciences, DOI: 10.1007/978-3-642-32521-2,
© The Author(s) 2013